Bryan Appleyard is a special feature writer for the *Sunday Times*. He specialises in writing about the relationship between science and popular culture and is author of numerous critically acclaimed books. He has twice won Feature Writer of the Year.

UNDERSTANDING THE PRESENT

An Alternative History of Science

BRYAN APPLEYARD

TAURISPARKE
PAPERBACKS

Published in 2004 by Tauris Parke Paperbacks
an imprint of I.B.Tauris & Co Ltd
6 Salem Road, London W2 4BU
175 Fifth Avenue, New York NY 10010
www.ibtauris.com

In the United States of America and in Canada distributed by
Palgrave Macmillan, a division of St Martin's Press
175 Fifth Avenue, New York NY 10010

First published in 1992 by Pan Books Ltd, a division of Pan Macmillan Ltd

Cover image: an Armilliary Sphere reproduced by kind permission of the
Science Museum/Science & Society Picture Library

ISBN 1 86064 891 6

A full CIP record for this book is available from the British Library
A full CIP record for this book is available from the Library of Congress

Library of Congress catalog card: available

Printed and bound in Great Britain by MPG Books Ltd, Bodmin

For Fiona Appleyard

*The more enlightened our houses are,
the more their walls ooze ghosts.*

ITALO CALVINO

*All history must be mobilized
if one would understand the present.*

FERNAND BRAUDEL

Contents

Preface   xi

Introduction   xvii

1   **Science works, but is it the truth?**   1

2   **The birth of science**   17

3   **The humbling of man**   48

4   **Defending the faith**   79

5   **From scientific horror to the green solution**   110

6   **A new strange mask for science**   138

7   **New wonders . . . new meanings**   168

8   **The assault on the self**   199

9   **The humbling of science**   227

Glossary   251

Notes   261

Bibliography   268

Index   273

Acknowledgements

I would like to thank the following for the time they allowed me to discuss the subject matter of this book:

Don Cupitt, Lecturer in the Philosophy of Religion at the University of Cambridge and Dean of Emmanuel College.

Richard Dawkins, Lecturer in Zoology at the University of Oxford.

John Durant, Visiting Professor of the History and Public Understanding of Science at Imperial College, London.

Professor Brian Hartley, director of the Centre for Biotechnology at Imperial College, London.

Roger Penrose, Rouse Ball Professor of Mathematics at the University of Oxford.

John Polkinghorne, President of Queens' College, Cambridge.

Michael Redhead, Professor of the History and Philosophy of Science at the University of Cambridge.

Stephen Hawking, Lucasian Professor of Mathematics at the University of Cambridge, contributed, perhaps unwittingly, to the genesis and content of this book when, in 1988, he granted me an interview as a journalist for *The Sunday Times Magazine* as did his wife, Jane Hawking.

I would also like to thank Peter Straus and Giles Gordon for their enthusiasm and support for this project. Many others, family and friends, have been punished over the years by the obsession that inspires this book. Thanking them might be inappropriate, but I certainly apologize. I hope they think it was worth it.

Preface

I wrote *Understanding the Present* just over a decade ago because contemporary science was running out of control. We were entering a new era of scientific triumphalism. In books, newspaper articles and on television extravagant claims were being made for science that were simply not true. Science is not a unique, omnipotent force that will, by itself, make a better world. It is not a truth about the world that renders all other truths obsolete. It is not on the verge of explaining what Douglas Adams called life, the universe and everything, nor will it ever be. And, finally, its past achievements have not proved unambiguously good for the human species or the planet.

I wished to counteract a new wave of scientistic propaganda by making two points: that the benefits of science are not cost-free and that we should not be deluded by those benefits into thinking that science can provide salvation from the human predicament. Both points seemed as self-evident to me then as they do now. But, just as they urgently needed to be made in 1992, so they need to be restated in 2004. If, to some, the arguments of *Understanding the Present* had once seemed abstract and remote, now they are urgent and personal. The extravagant claims are still being made. Incredibly, after a decade in which cloning, the uncovering of the vast Soviet biological weapons programme, the mysteries of BSE, the appalling damage being done to the environment, the irreducible complexity of the post Cold War world and the advent of the technologically adept terrorist have vividly exposed the deep ambiguity of science's role and its ultimate inability to solve the most important problems we face, there are still high-profile scientists and philosophers preaching that we shall be saved if – and only if – we put our faith in science.

This book was never intended to be anti-science, a Luddite

tract, and neither was it intended to deny the creative and intellectual power of science. Science always thrilled me, as it must anybody possessed of an imagination. But it thrilled me as one human activity among others. The fantastic world revealed by quantum theory overturns our common sense apprehension of nature; but so does the Cezanne *Still Life*, painted in 1900, the same year that Max Planck took his walk in the Grunewald woods. And there is certainly as much grandeur in the great edifice of Christianity or the serene wisdom of Buddhism as there is in the history of life outlined in Charles Darwin's *Origin of Species*.

The insight that I had begun to glimpse in the late eighties and which crystallized into *Understanding the Present* was that influential people – and the people they influenced – could no longer be relied upon to accept such a balanced account of human affairs. Scientism, the extremist belief in the omnicompetence of science, had re-emerged. I was astounded to hear some of the claims being made and the rhetoric employed; astounded, first, because previous scientistic thought had been so thoroughly and eloquently discredited and, secondly, because Nazism and Communism, the two great disasters of the twentieth century and, in terms of body counts, the two greatest disasters of all time, both had explicitly scientistic roots. Surely, we should have learned that such a reduced idea of human wisdom could form no basis for the organization of society. Science and technology had been shown to be as much tainted with original sin as the people that made them.

After publication, however, it became even more apparent that scientism was on the march. Famous figures emerged to denounce me as a heretic. To question any aspect of science, it seemed, constituted thought crime. I do not, with that phrase, evoke George Orwell's novel *1984* lightly – at the time I felt exactly like its tortured protagonist, Winston Smith, a man confronted by a baffling, implacable and patently false ideology. Why, I wondered, were they so angry?

The answer was obvious. The scientistic idea – that humans, alone among all species, can fundamentally improve and, indeed, perfect ourselves and the world - is not rational and is certainly not supported by scientific evidence. It is a faith like any other.

And, like other faiths, it has its fundamentalist adherents who feel that any threat to its key doctrines must be crushed. The fact that this faith considered itself to have discovered a perfect rationality is neither here nor there. All faiths are grounded on this conviction.

The virulence of these responses made my case. To argue, as one scientist did, that this book convinced him that pornography was to sex what philosophy was to science is to admit that you have abandoned literacy and wisdom in favour of a peculiarly vulgar megalomania. Scientism was not just on the march, it was snarling and hungry for conquest. Of course, there were other, more reasoned and sensitive voices raised in my defence and to those voices I was and am grateful. But, from the scientists who had access to the media, the tone was a uniform howl of outrage. Years later I was introduced to one of these howlers. 'I despise you,' he said to me. And that's all he said.

Then, slowly, things began to change. The Theory of Everything, which had been widely promised by physicists within a few years if not months in 1992, seemed to be receding into the more distant future. Research into artifical intelligence, the Holy Grail of computer science, had ground to a rancorous halt. I noticed the appearance of a slightly more sceptical tone in press coverage of the wilder dreams of physicists and a greater articulacy among my fellow opponents of scientism in public debates and lectures. Intense radio discussions about the inner nature of science – rather than just bland celebrations of its achievements – became commonplace.

Arguments that seemed to have been buried by the scientistic consensus began to re-emerge in unexpected places. A discussion of my book appeared, bizarrely, in an Anne Rice novel – *Memnoch the Devil* – and the late Carl Sagan gave me an unusually fair hearing in his book *Pale Blue Dot*. But it was developments in biology that did most to quell the scientistic rhetoric of the late eighties and early nineties. The deciphering of the DNA molecule in 1953 had marked the moment when biology took over from physics as the most imaginatively resonant of all the sciences. This change did not fully become apparent to lay people until the nineties for it was then that practical applications of this new knowledge began to appear either as realities or as imminent possibilities –

designing babies, bio-engineering crops and farm animals, pre-natally detecting genetic abnormalities and, with the appearance of Dolly the sheep, cloning, animal and human. Then, on 11th September 2001, we all became aware of the threat of inter-national terrorism. That low tech event drew everybody's atten-tion to the high tech event that might follow – a biological attack. The secret Soviet weapons programme has left behind vast stocks of lethal pathogens, most alarmingly bio-engineered smallpox. We do not know where they have gone. Let loose on an un-vaccinated world, smallpox could kill millions. Controlling the activities of scientists was no longer a moral or political necessity, it was a matter of the survival of human civilization.

Suddenly, lay debate about the inroads of science into our lives became urgent and essential and, equally suddenly, the hard scientism of the early nineties looked very hollow indeed, if not positively wicked. Biology was making intimate and momentous claims about human life, claims that could not be blandly sub-sumed under the heading of 'progress'. It was also threatening to engineer novel bugs, as if there weren't enough to worry about. Even if the bio-weapons threat never materialised, the nightmare world of total reproductive control prophesied by Aldous Huxley in his novel *Brave New World* seemed to be just around the corner.

Welcome as the debates about biology were, there was a problem. Intellectually, bio-sceptics seemed to be trapped by the terms of secular humanism. Repeatedly, the debates declined into competing versions of simple utilitarianism or consequentialism. So, for example, sceptics argued we should not allow human cloning or use genetically modified organisms merely because they would have dire outcomes. The weakness of this position is that, if they can be shown not to have such outcomes, then there could be no reason not to go ahead. What was missing was a con-fident statement either that these things should not be done because they were wrong in themselves or a sophisticated con-sequentialist assertion that they should not be done because cal-culating the outcomes was beyond human competence. In other words, both the sceptical and the scientistic were labouring under the same delusion. Once the details and the surplus rhetorical noise are stripped away, that delusion is perfectibility, the belief that humans have a unique power to take complete control of themselves and the world.

Understanding the Present anticipates this problem as, in fiction, does Huxley's novel. In *Brave New World* the central confrontation is between the scientistic view of life that says if, artificially, you can be made happy, then there can be no reason why you should not embrace that possibility and the sceptical view that such a condition is shallow and empty. Why read Shakespeare's *Othello*, asks Huxley's mandarin, when the play's issues have been made meaningless by drugs and reproductive control?

Understanding the Present attempts to provide an answer primarily by pointing out the disjunction between the lived experience of our lives and the visions of science and by insisting that the latter must not – cannot – persuade us to abandon our faith in the former, whether that faith is religious, cultural or intuitive. The failure of something else that happened in the nineties – globalization and its cult of the free market – should have demonstrated clearly enough that cherished ways of life will not be crushed simply because one particular way of life believes it has a unique claim on rationality.

Finally, though I would not wish to change the text, I would like in this introduction to modify slightly the argument of this book in the light of my own experience of the debates of the last decade. The pre-eminence of biology in these debates has led me to think more deeply about the scientist whose mighty ghost haunts our age – Charles Darwin. It has also led me to think more seriously about a cause to which I give excessively short shrift in *Understanding the Present* – environmentalism.

The secular humanist interpretation of Darwin is that, in explaining our presence in the world, he justifies our conviction that we are uniquely able to take control. This bizarre and, to me, evidently mistaken interpretation surfaces again and again among Neo-Darwinians and it even appears as the one flaw in the work of that great biologist and environmentalist Edward O. Wilson. In fact, the only credible interpretation of the Darwinian insight is that it denies rather than affirms our uniqueness and, therefore, our notions of perfectibility. In short, it should humble us and teach us to be wary of our hubristic delusions of competence.

At its highest levels – I am thinking here of Wilson at his best but, most importantly, of James Lovelock – contemporary environmentalism represents an understanding of this great post-Darwinian truth. To the human imagination, such thinkers offer a way out of the arid and destructive scientism that remains the default mode of much contemporary discourse. This way out is not an answer – humanity came up with plenty of answers in the twentieth century, all of them catastrophic – but it is a way of changing if not minds then, perhaps, souls, of rediscovering the transcendent virtue of humility.

Understanding the Present aspires, from outside science, to do the same. It hopes to place us back in the real world in which, for good or ill, we live and back, above all, in the present rather than in the illusory future, scientism's land of infinite promise that will always, thanks to our fallen condition, elude our grasp.

Introduction

I grew up in the shadow of science. I used to think that this meant simply that I came from a scientific family. I remember when I was perhaps seven or eight asking my father, an engineer, how much water could be held by a water-tower close to our home. He worked it out on the spot. I was dumbstruck and made uneasy by this power. It inspired me to imitate him by developing a facility for mental arithmetic which I would use to impress other children.

On my father's side there had been an aunt and an uncle, both of legendary intelligence and both romantically fated to die too young for me to meet them. They were physicists, a word that possessed as much magic for me as 'sorcerer' or 'priest'. It was assumed, reasonably enough given my early enthusiasms, that I would follow in their footsteps. But I was born in 1951 and so I rebelled, as most of my generation did, in the conviction that to be different from my parents was my particular obligation and destiny.

I became self-consciously aesthetic, insisting on the superior virtues of unscientific creativity and using this to oppress my brother who *had* followed in the pre-ordained footsteps. But, as time went by and I wrote and thought more, the language of this artiness came to seem oddly thin. I felt I had come almost to the end of all these critical formulations and assertions. And they appeared, suddenly, groundless and insubstantial. The preoccupations of those who talked about art, if not of the artists themselves, began to seem hopelessly trivial. They were talking to each other in a tiny room while ignoring something vast and terribly relevant that was happening outside.

While writing my book *The Pleasure of Peace* about the arts in postwar Britain, I began to turn back to science. At first I did so as if in defiance of the artists and critics. What did they know?

Here were *facts*. And they were splendid facts, gigantic facts encompassing all of time and space. The travels of contemporary artists began to seem like a drive round the block compared to these fabulous voyages. This, I concluded, was the great truth of the modern world that these precious artists were so studiously ignoring.

Perhaps I was simply recapturing my childhood. Certainly I was recapturing old habits, for now I began to acquire a facility with the concepts of science and to use them to humble my artistic coevals.

But I had been uneasy with my father's ability to measure the volume of water in the tower. I sensed something dangerous and ominous in this strange wisdom. As I read and talked more of science and looked at the contemporary world, I began to see the point of my infantile unease. I began to see that I was not the only one to have grown up in the shadow of science.

I wrote this book in the conviction that science, more than anything else, has made us who we are; science is our faith and our age's unique signature. My conclusion is equally simple: we must resist and the time to do so is now.

Less simple is the task of persuading you either of my thesis or my conclusion. I feel like the hero of the film *Invasion of the Body Snatchers*, racing about town warning the sceptical that our neighbours are in reality, alien invaders in disguise. There is much that stands in the way of acceptance of what this madman believes to be the truth because science, like the aliens that look and live like humans, is *buried* within us, it is concealed. In order to expose its workings we need to look beneath the fabric of contemporary life. This involves understanding the past, something science itself seldom needs to do, and it requires an effort of the imagination to see how we have been formed by the struggles of the last 400 years if we are to confront the demon of our new form of knowledge.

Scientists themselves are of surprisingly little help. They find it difficult to talk of what they do because they tend to assume detailed knowledge is required for generalities to be understood. They find it hard to grasp the concept of the *meaning* of their work, assuming this to be a debate that takes place at a lower level than the specialized discussions with their colleagues. When they do generalize – or 'popularize' as it is usually called with a noticeable degree of contempt – they tend to reveal a startling philosophical naïvety.

I once spoke to the physicist Stephen Hawking. In reply to one of my questions he quoted and, I believe, fundamentally misunderstood a remark of the philosopher Ludwig Wittgenstein. Hawking repeats the quotation in his book – 'The sole remaining task for philosophy is the analysis of language.' [1] and derides the sentiment – 'What a comedown . . . !' I attempted to correct him, as I shall show later in this book Wittgenstein's insight has immense and profound implications. But Hawking simply would not listen. 'I do not think so,' was his only response.

This is a dangerous state of affairs. Scientists need to be observed and criticized more than any other members of society. I say this not just because of the horrors that might emerge from their laboratories, but also because of the necessity for making them as morally and philosophically answerable as the rest of us. This is the reason, above all others, why this particular history of science can only be written by a non-scientist. All accounts from within the temple of science are fatally flawed by the element, often unconscious, of defensive propaganda.

So this book is, first of all, a history of science and of our attempts – intellectual, spiritual and moral – to cope with its burgeoning power and effectiveness. Many will see it as a highly biased history. I make no apologies for that. As I shall show, bias – very distinguished bias – has also been busily at work on the other side of this argument.

Understanding the present requires us to make this effort to dig beneath the surfaces of appearances and daily life to find the ideas, prejudices and restrictions that form this life. Conventionally if asked to describe our world or our age we talk of politics, economics, sociology, even anthropology as if attempting to produce an explicit, visible image of the way things work or appear. But this is description. Understanding requires a grasp of the dynamic that made all these politics, these economics and this sociology. In other civilizations such a dynamic may be religious or some practical demand for survival. In ours, I am convinced, it is scientific.

Understanding the present requires, therefore, an understanding of science. This does not mean that we have to be able to read pages of equations or even to grasp specific theories in detail. Much more important is the need to understand what science has been, what it now is and what it *means*. I must also show that this is only one meaning among many others available to us and yet it is the one that is threatening to conquer all the others. If I can show you all that, my thesis will become acceptable.

Persuading you to accept my conclusion requires something more. It requires a demonstration of what science can now do and is likely soon to be able to do. It requires an understanding of the appalling spiritual damage that science has done and how much more it can still do. Science, quietly and inexplicitly, is talking us into abandoning ourselves, our true selves. It is doing so today more effectively then ever before. That is why now is the time to resist.

The book is arranged as follows. The first chapter raises in embryonic form all the issues of the book by describing the effectiveness of science in our world and covering some of the ways science is habitually portrayed and celebrated. The second chapter begins my history with the foundations of the way of knowing that we now call science. Chapter Three deals with some of the philosophical implications of the new science of Galileo and Newton and takes the story up to the spiritual shock of Darwinism and the tragic vision of Freud. Chapter Four explains the religious and moral response to the new world created by science. Chapter Five turns to a specifically modern response: the dismay felt in the twentieth century at the more horrific inventions of science and the sudden and explosive growth of environmentalism as a way of defending ourselves against its impact. Chapter Six raises the issue of whether scientific developments of the twentieth century like quantum theory, relativity and chaos theory have fundamentally changed the nature of science and Chapter Seven discusses the numerous attempts to prove that they have done so – that somewhere in our new science we can find God or a different way of life. Chapter Eight concerns the last frontier which science is now approaching – the frontier of the human self. Finally, Chapter Nine draws together these themes and describes what I believe to be the only possible answer to the restless and impatient demands of science.

1
Science works, but is it the truth?

The future belongs to science
and those who make friends with science

Nehru[1]

At the conclusion of his book *A Brief History of Time*, Stephen Hawking discusses the possibility of an end to physics. This end would be a complete theory which unified all of space and time – a Theory of Everything. It would, initially at least, be no more than a set of equations. But these equations would implicitly contain everything that had happened or could ever happen. They would be the rules by which the game of existence is played. They would provide a mathematical model for the entire history of the universe. Applied with superhuman patience and determination, they could predict that a particular snowflake would fall on a particular blade of grass or that you would be reading this now.

Hawking holds out the hope that such a theory would, in time, become understandable to everybody. Clearly its mathematical detail would still be understood by only a few. But its broad principles would slowly permeate our culture. This has happened before: most of us can now grasp the physics of Isaac Newton without understanding his mathematics and even Albert Einstein's Theory of Relativity is gradually becoming a familiar part of the modern picture. So the Theory of Everything would become part of our lives.

'Then,' Hawking writes, 'we shall all, philosophers, scientists, and just ordinary people, be able to take part in the discussion of why it is that we and the universe exist. If we find the answer to that, it would be the ultimate triumph of human reason – for then we should know the mind of God.'[2]

Hawking's tone and his conception of the significance of his work are typical of a certain way of presenting science. Almost all popularizers of science – notably, in recent years, Jacob Bronowski and Carl Sagan – say the same kind of things. They say that science is a spectacle of majestic progression, that, in spite of its apparent obscurity, it is a natural and inevitable product of the human imagination, it has fundamental human significance and it is ultimately capable of answering every question.

God is often evoked. Sagan in his introduction to Hawking's book says: 'This is also a book about God . . . or perhaps about the absence of God. The word God fills these pages.'[3] Bringing God into the equations suggests both the importance and virtue of the scientific enterprise – this, we are being told, is a continuation of the ancient religious quest to find Him and to do His will.

The message is that science is *the* human project. It is what we are intended to do. It is the only adventure. Bronowski, in particular, presents science as that which has always made us distinctively human. Science and technology accompany all human societies and distinguish us from the beasts. They are continuous throughout history: relativity and microwave ovens are clearly the descendants of the first plough or the first wheel; they spring from the same impulse, the same inspiration. Most persuasive of all, ploughs and microwaves are unique in the known universe in that they are fashioned by reason.

This is propaganda, dangerously seductive propaganda. It is all misleading, even offensive, to the lives we actually lead. We are diminished by this rhetoric. It is the rhetoric of what is sometimes called 'scientism' – the belief that science is or can be the complete and only explanation. An important part of any case is that, whether we or more modest scientists like it or not, science possesses an intrinsically domineering quality. This kind of triumphant scientism is built into all science. Opposition tends to be subdued and demoralized to the point where we can no longer identify the damage done by these popularizers.

The appearance of a Hawking, a Sagan or a Bronowski in the

bestseller lists or on television may be a huge media event, but it is quite rare. Every decade or so we seem to be ready for a new popularizing figure to bring us news from the further reaches of speculative and theoretical science or to encourage our faith in its virtue. In the intervening periods science blends innocuously into the background noise of our culture. It is celebrated in chatty television or magazine items about this or that invention or innovation. But these are space-fillers between what we think is the important news of politics, economics or the dramas of human relationships.

The word 'science' itself almost vanishes. When used it may dimly evoke images of schoolrooms, laboratories or men in white coats, a rocket launch, a nuclear explosion or a chemical plant. We may see equations, computers, test-tubes, particle accelerators or colourful, toy-like models of molecules. Or the word may evoke technology: televisions, cars, manufacturing techniques, building methods, communications systems. If pressed, we may bring ourselves to acknowledge that, in the developed world, we cannot dress, feed, travel, procreate or be entertained without the intervention of science. But we tend to think these are all different things. The electric kettle is not the same as an aircraft. They are both machines, certainly, but that is all. So our conception of science is diluted and its true identity concealed. For science is one thing and it is in both the kettle and the plane.

But, subliminally, our vague awareness of and gratitude for the ease and ubiquity of technology prepares us to accept the larger claim of science that it alone can lead us to God. For we can see all about us how much science can do; perhaps it can do this as well. It has solved so many of our little problems, maybe it can solve the big one. After all, both flying and electrically boiling water are miraculous in their different ways and our idea of God is usually accompanied by miracles.

This unarguable and spectacular effectiveness is the ace up science's sleeve. Whatever else we may think of it, we have to accept that science works. Penicillin cures disease, aircraft fly, crops grow more intensively because of fertilizers, and so on.

'It is science alone', said Jawaharlal Nehru, the first Prime Minister of India after the departure of the British colonists, 'that can solve the problems of hunger and poverty, of insanitation and illiteracy, of superstition and deadening custom and tradition, of vast resources running to waste, of a rich country inhabited by starving people . . . Who indeed could afford to ignore science today? At every turn we have to seek its aid . . . The future belongs to science and those who make friends with science.'

All attempts to understand the present have to begin with this modern conviction that for every problem there is a scientific solution.

'The priest persuades humble people to endure their hard lot,' writes the molecular biologist Max Perutz, 'the politician urges them to rebel against it; and the scientist thinks of a method that does away with the hard lot altogether.'[4]

Science tells us that there are things called problems that have things called solutions and it tells us by showing us. You are dissatisfied with the quality and convenience of the music in your home? Here is a compact disc player. You wish to avoid small-pox? Here is an injection. You wish to go to the moon? Here is a rocket. You are hungry? Here is how to grow more food. You are too fat? Here is how to lose weight. You feel bad? Here is a pill, feel better. No problem, says science.

We are so used to this idea that we forget how new it is in human history. It is perhaps too banal to point out that the Ancient Egyptians died of smallpox because they did not have vaccines and neither did they have electronically reproduced music in their homes. But it is not banal to note that they did not have *any* of these things. Science and technology have not developed gradually over the whole history of human culture; they have suddenly exploded all about us. Their sheer, profligate effectiveness is something utterly novel.

Bronowski would deny this, claiming that science and technology have suddenly become so successful because human reason attained a kind of evolutionary take-off point. Nothing different happened, merely more of the same inspiration that invented the wheel. It is central to my thesis that this is wrong, that science is a fundamentally new way of knowing and doing things. I believe that an examination of scientific history makes this point obvious. I find it absurd, almost sentimental, to say, as would Bronowski, that a plough is like a CD-player. They are fundamentally different. The designer of the latter has to have a different way of knowing from the maker of the former. To deny this, I hope I will show, is to drain all meaning from the words 'new' and 'different'.

But, for the moment, let me stay with the most seductive aspect of science – its practical effectiveness. In boiling our water or staving off disease, science is showing us that we – the human race – can do extraordinary things that are both miraculous and superbly useful. This is consoling; it suggests we need not be passive victims, mere observers of our destiny. We can improve on what we have been given by nature. But this improvement is not a simple, innocent matter of doing things better; it has certain implications.

Take the idea of the map. We use maps all the time and we think nothing of how they work. Our modern maps are complete and clear; there is nothing missing and there is nothing we cannot understand. Old maps show some regions with a reasonable degree of certainty. But then knowledge fails and the imagination of the mapmaker takes over. The region of the known shades away into myths and fairy stories – dragons and giants at the world's end, a landscape of chaos beyond the limits of order. There was a line drawn to mark the limits of human knowledge. There was an outside, a beyond.

Now our maps are complete, but not because we have been everywhere and seen everything. Our maps are complete because we have found *a better way of making them* that excludes the need for dragons. Indeed, the golden key to the success of science is precisely captured by the realization that we can map places without visiting them. By drawing lines of longitude and latitude and by astronomical observation we can produce an effective picture of the whole world.

Imagine you are a traveller looking round an alien landscape. There are trees, rivers and mountains. But they are meaningless in themselves. You cannot say where you are simply because of that mountain or this tree. You can spend your days finding out everything about what you see, but it will never tell you where you are. But if I give you an effective map with your mountain and your river marked upon it, the world is transformed. You can calculate your position relative to all other positions. From any point you can, therefore, journey to any other point. This is not simply better knowledge, it is utterly different knowledge.

And, once we have such a map, the old map with its unknown regions immediately becomes naïve. The new map tells us that nothing can be ultimately unknown. We may not know precisely what we may find when we arrive anywhere, but we do know that the place exists in this position relative to this sea, island or mountain or to our home. The wisdom of the past has become quaint because it is ineffective. Modern man with his maps is infinitely more powerful than those poor people who thought there were dragons at the world's end. He is like a god.

One way of knowing – the casting of an invisible net of latitude and longitude over the earth's surface – has proved spectacularly effective. It has convinced us of our power. Civilized man with his map can travel far from civilization, secure in the conviction that he is tied to his home by an unbreakable chain of knowledge and calculation. And, once these lines are secure, they are followed by others – telecommunication cables, radio and microwave links. Finally we have our complete modern, mental map of the world: a blue-green globe nestling in an invisible field of voices and crossed by rapid beeps and streams of computerized information. We have killed the dragons.

So the effectiveness of science gives us more than hot water or the facility to hear good music, it gives us a sense that we can grasp everything, even things we cannot see. Our maps convince us that, on one level at least, our world is now completely known. We are superior to a primitive community without such effective and complete maps because we can point to the place where they are and how far they are from every other place. They can only construct myths about the land they cannot see.

This illustrates an important, higher aspect of science's effectiveness. It shows that mere technology is only the most obvious demonstration of our new powers. For, above the gadgets, there is the sense that science provides us with a way of doing and knowing almost anything we like. By providing its maps, its nets of explanation, it convinces us that what is unknown can only be a question of detail. We may not know if there is a tree at this particular grid reference, but it is only a question of finding out, of sketching in the details. The map has let us control the space in our minds without having to see it with our eyes.

This higher type of effectiveness gives us an immense confidence in our powers. We feel that problems always have solutions waiting for them at some point in the future. We know the techniques, the territories in which these problem are to be found; it is merely a question of a closer reading of the right maps. For example, we know, almost to an inch, where the planet Mars is to be found; flying there is merely a matter of technological detail. Conceptually we have already landed and conquered.

So science is technologically and conceptually effective. It breeds total confidence. It is unarguable, even if you reject the culture from which it springs. Iraq's President Saddam Hussein may have led his country into battle with evocations of an Islamic order bitterly opposed to the West, yet he knew that he required the knowledge of the West to guide his missiles and to build his bombs. He required science's rigorous power and simplicity. Once an enemy is simplified to a point on a map by computers and satellites, he can be destroyed. A problem is solved. And Christians and Muslims alike are killed by scientific high explosive. Science unifies our ambitions, simplifies our attainment of them and, thereby, grants us power beyond our dreams.

Saddam Hussein's two-faced posture points to a further essential insight that leads us to an understanding of science and of the present. For science is exclusively the product of the West – by which I mean the European-American culture. No other culture has produced either the same science or any other equally effective form of knowledge. Certainly there are other forms of wisdom, but none that have discovered this particular type of effectiveness. Like Saddam, many have wished to distance themselves from the West while benefiting from its science. They wish to do so because of the obvious power granted by science. No Buddhist, Muslim or Confucian culture has produced a better way of increasing crop yields, curing disease or killing people than the Western way. So those cultures, when they wish to eat, live or fight, turn to the West's science.

In doing so they make the assumption that science will be a neutral import. They can remain good Buddhists or whatever.

No Islamic state, for example, feels its religion is threatened by the Western technology employed to extract its oil. The Christians happen to have the machines and the Muslims happen to have the oil. This is of no transcendent significance.

The assumption here is that science's levels of effectiveness can be disconnected from any higher meaning. It works, it gives power and control. But it does not tell us anything of absolute significance about ultimate reality, so it need not affront our deeper beliefs. The Koran is the book of truth; science is merely the book of how to do things.

This is an illusion on at least two levels. First, the practical effect of science in the world is certainly not neutral. Secondly, there is the weakness of the pious hope that science and religion are independent realms which can easily be separated. It is a central contention of this book that our science, as it is now, is absolutely not compatible with religion.

Science has been seen as a morally neutral, practical benefit, that any culture can employ without danger of corruption or contamination. The practical impact of science on non-Western cultures is obvious; the pattern is absurdly familiar. Say we have a sick child of an isolated people. He can be cured with penicillin or he can die. A Western doctor gives him penicillin and he lives. The people decide they want more penicillin and then other drugs. Soon they want other goods as well, some because they are life-saving like the medicines and some because they are just nice. To buy the goods they must trade. Trade draws them into the economic system of the developed world and so on. This process is universal and probably irreversible. It smooths out local cultural differences and unifies the ways in which human life is conducted. We all want penicillin and we all must pay for it in roughly the same way.

If we feel protective towards those primitive people, we have a choice. We can deny them the penicillin and let the child die in the name of the long-term autonomy of his people's culture. Or we can give him the drug and hope the culture can still be saved.

But that hope is usually no more than pious. When local wisdom is humiliated by science and local culture is drawn into scientific civilization, it is difficult to believe that what remains can continue to be a unique culture in any meaningful sense. Rather it becomes a self-consciously protected museum piece or, most saddeningly, a source of fashion accessories. We collect the people's artefacts and learnedly discuss their significance. Or we dress in 'ethnic' clothes in celebration of, for example, the American Indian. But the truth is that all the artefacts, clothes and rituals were diminished at the moment the penicillin was first administered. Their absolute meaning was made relative by the devastating effectiveness of science.

The point to remember here is that this effectiveness is absolute. Science transports the entire issue of life on earth from the realm of the moral or the transcendent to the realm of the feasible. This child *can* be cured, this bomb *can* be dropped. 'Can' supersedes 'should'; 'ability' supersedes 'obligation'; 'No problem!' supersedes 'love'.

Science is not a neutral or innocent commodity which can be employed as a convenience by people wishing to partake only of the West's material power. Rather it is spiritually corrosive, burning away ancient authorities and traditions. It cannot really co-exist with anything. Scientists inevitably take on the mantle of the wizards, sorcerers and witch-doctors. Their miracle cures are our spells, their experiments our rituals.

So, as it burns away all competition, the question becomes: what kind of life is it that science offers to its people? How does it replace other wisdom, other meanings? These are the questions of the nature of the scientific life in the scientific society and they are the questions that will lead us inexorably back to Hawking's God.

Science is effective, but what does it tell us about ourselves and how we must live? The brief answer to this is: nothing. Science has always worked assiduously to avoid being a religion, faith or morality. It does not tell us why we should do things or how we should live; it offers, instead, solutions. Life is a series of separate problems with separate answers. It is not an issue in itself so much as a container of issues.

The primary characteristic of these discrete solutions is that they are not committed or conclusive. They work for the moment. They can be changed if something better comes along. Communism missed this point. It was supposedly a scientific way of organizing society. But Marx and Lenin naïvely took science to mean an absolutely final system that would explain all history and to which it was simply necessary to subject oneself. But the only way for science to be effective as a maker of societies is for it to resist any final commitment to a particular solution. It is possible to be right, but not *more* right. It is, however, always possible to be *more* effective.

Modern Western society, which I shall call 'liberal' society (see the Glossary on page 251 for a justification of my usage of this and other contentious words), is a realization of this scientific method. Government is neutral. It provides a secure arena of law and order in which, within certain limits, people can pursue their own beliefs. Neither government nor society as a whole provides moral direction or meaning, rather it simply safeguards tolerance so that individuals or groups can pursue their own meanings. So a modern democracy can be expected to include a number of contradictory religious faiths which are obliged to agree on a certain limited number of general injunctions, but no more. They must not burn each other's places of worship, but they may deny, even abuse each other's God. This is the effective, scientific way of proceeding.

In a recent book – *The End of History and the Last Man* – the American Francis Fukuyama has put forward the theory that liberal democracy is the culmination of the historical process. Such societies have removed themselves from history in the sense that they are no longer subject to fundamental ideological debates about the best form of social and political organization. Science is also the key to his analysis. Fukuyama believes that science, because of the way it cannot be unlearned and its method of building on each successive achievement, introduced for the first time a clear direction into history. His optimism about this process is tempered by his own uncertainty about where liberal-democratic-scientific man is to find a purpose in life.

For, as I said, science cannot co-exist. This is not just true when it is being exported from one nation to another, but also when it competes with other systems within a single nation. The science-based liberal democracies, therefore, tend towards a unity of unbelief.

This, I know, is contentious. In Western Europe religious

belief appears to be declining in both numbers and effectiveness, but it is strong in the United States and, in Eastern Europe, religious feeling was a crucial element in the overthrow of communism. There are contradictory figures showing that religious observance is both increasing and decreasing in almost every country. My assertion, therefore, is an opinion but it is one that is supported by all my experience of these countries.

In any case, the point is not absolutely central. For it is still possible to provide some generalizations about the spiritual mood of these developed, scientific nations even if their particular religious temper is uncertain. Science is perfectly capable of marginalizing believers without actually stripping them of their belief.

This spiritual mood arises from the enforced neutrality of scientific liberalism that I have described. To sustain its effectiveness science insists upon a universally open-ended view of the world that accepts and embraces the permanent possibility of change and progress. At any one time scientific man can only regard his knowledge as provisional because something more effective might come along. He may construct private absolutes of faith or morality, but, in public, he must inhabit a fluid, relative world. So, for example, his moral choices cannot be made by referring to an outside order or system, they can only be *his* choices. Given that, he will always be aware that there are different choices made by other people. He cannot argue absolutely against these different choices, he can only say that he thinks they are wrong. And even this degree of conviction must be open to change. There is only relative right or wrongness; there is no absolute form of either.

The same process applies to his spiritual condition. He may define his own identity and purpose in life, but he must do so in the knowledge that other identities and purposes have been chosen and must be respected as equally valid as long as they conform to the broad norms of behaviour of the liberal society. He cannot even tell his children with any conviction that they must believe what he does because it is true. They can simply point out that he is offering them not a fact but just another opinion among countless others.

The obvious point about this is that it makes it progressively more difficult to sustain either a morality or a spiritual conviction. Daily, liberal-scientific man is made aware of the arbitrariness of the exercise. He finds he cannot even defend his position because all arguments end in the blank inconclusiveness of the total mutual tolerance that is the one thing required of the combatants. 'We agree to differ' is the standard form towards which all conflicts in a liberal society tend. And, finally, because of the aridity of such a conclusion, even the energy to differ expires. Unable to create a solidity for himself, liberal man lapses into a form of spiritual fatigue, a state of apathy in which he decides such wider, grander questions are hardly worth addressing. The symptoms of this lethargy are all about us. The pessimism, anguish, scepticism and despair of so much twentieth-century art and literature are expressions of the fact that there is nothing 'big' worth talking about any more, there is no meaning to be elucidated.

In this condition liberal-scientific man is, of course, in no fit state to resist the type of answers offered by the scientist. He finds himself waiting, as an ordinary man, for the coming of Hawking's God.

It is tempting for liberal apologists to see this bewildered condition not as a specific state, but rather as the normal human destiny. All men, say the liberals writing their own history backwards in time, have suffered this bewilderment. But, even in recent history, it is clear that this is absurd. In the clash between fundamentalist Islam and the liberal West, the reality of liberalism as a distinct attitude, a cultural decision, becomes obvious. In Iran, Iraq, Afghanistan and in the internal politics of almost all Arab countries, violent tensions have been created by the upsurge of a form of Islam that finds the tolerant co-existence of religious beliefs incomprehensible, indeed threatening to the process of salvation.

Most pointedly there was the Salman Rushdie affair in which an Anglo-Indian novelist was sentenced to death by the Iranian Government for writing a book – *The Satanic Verses* – which was interpreted as an offence against the faith. The fact that one country saw itself as justified in sentencing the citizen of another to death bewildered liberal Westerners. Not only did freedom of speech appear to mean nothing to these people, they even regarded the sovereignty of the nation state as secondary to the demands of their religion.

But the post-revolutionary Iranian state is founded upon the conviction that Islam is absolutely true and universally applicable. And if that religion be true, incontrovertibly true, then why should the Iranians not feel justified? As the Iranian ambassador to the Vatican said on the day after Rushdie had been sentenced by the Ayatollah Khomeini: 'Why do you find this behaviour strange?'

A liberal answer to the question could only be feebly pragmatic: we find it strange because we have found tolerance a more practical virtue than vengeance upon the infidel. But, the ambassador could reply, what is the virtue in tolerance if we have found the truth? And surely to be religious at all means that you have found the truth. Liberals can only pretend to have private truths because they are not prepared to back them with their lives.

This leads on to the second point about the spiritual condition of scientific liberalism. Because it offers no truth, no guiding light and no path, it can tell the individual nothing about his place or purpose in the world. In practice this is seen as liberalism's great, shining virtue, for it is the one way of avoiding what the liberal sees as the horrors of the past.

Liberal history says that societies that did tell the individual who he was, what he was for and precisely how he should behave have almost invariably been cruel and destructive. Nazi Germany and Stalinist Russia were the great recent European examples. But, before the triumph of enlightenment and liberalism, it can be said that virtually all societies suffered from the vice of institutionalized intolerance. People suffered and died for their national religious or moral differences. Yet these differences are natural, they are the fundamental aspect of the human picture. Liberalism, institutionalized tolerance, would seem to be the only way of constructing a stable society that would sustain rather than oppress such a healthy plurality.

This is the key defence of liberalism's refusal to be spiritually committed. It says simply that this society has its faults, but it is the best we have evolved so far. 'It has been said', said Sir Winston Churchill in the House of Commons in 1947, 'that

13

Democracy is the worst form of government except all those other forms that have been tried from time to time.'

This is the utterly sound scientific defence of the scientific society. Just as we may say we do not know everything about the nature of matter, yet we can make an effective laser beam, so we can say that, though we do not fully understand society and are prepared for a better idea to come along, this liberal democracy is the most effective system we have yet produced. It copes with diversity so long as the constituents of that diversity are prepared to acquiesce.

But, sound as that defence is, it does not end the debate. For, in order to achieve this degree of effectiveness, the liberal state has had to abandon the role of the state as spiritual provider and so leave the individual in the bewildered quandary I have described. Perhaps it is better to be bewildered, alive and rich than to be certain, dead and poor. But that cannot be the end of the matter. For, as I have said, science is not neutral, it invades any private certainties we may establish as a defence against the bland noncommittal world of liberalism. It saps our energy. There must be a real danger that liberal society, having triumphed economically and politically, may now decline beneath the weight of its own spiritual indecision.

Fukuyama may be premature in announcing the end of history because this spiritual vacuum may be the fatally unstable element in the liberal democracies.

Tolerance becomes apathy because tolerance in itself does not logically represent a positive virtue or goal. So the tolerant society can easily decline into a society that cares nothing for its own sustenance and continuity. The fact that the democracies constantly seem to have a crisis in their schools is important – it is a symptom of a crucial uncertainty about what there is to teach, about whether there is anything to teach.

At the heart of this spiritual problem lies the lack of a sense of self. Just as scientific liberalism holds back from the moral or the transcendent, so it also holds back from providing the individual with an awareness of his place in the world. On the maps provided by science we find everything except ourselves.

This exclusion of the self from the explanations of science is a complex and profound matter that has implications that will surface again and again in this book. Here I will simply say that it cuts scientific man adrift from his moorings. Artistic expression over the past 400 years, the age of science, persistently returns to the man alone, lost and searching for something, though he is seldom sure precisely what. Even the devout Christian imagination felt the pressure of this uncertainty. The French mathematician and philosopher Blaise Pascal wrote in 1660, half a century after the date, 1609, I will choose for the start of the modern age, of the way we seemed to be able to find no way of establishing our place with certainty:

For in fact what is man in nature? A Nothing in comparison with the Infinite, an All in comparison with the Nothing, a mean between nothing and everything. Since he is infinitely removed from comprehending the extremes, the end of things and their beginning are hopelessly hidden from him in an impenetrable secret; he is equally incapable of seeing the Nothing from which he was made, and the Infinite in which he is swallowed up.

What will he do then, but perceive the appearance of the middle of things, in an eternal despair of knowing either their beginning or their end? All things proceed from the Nothing, and are borne towards the Infinite. Who will follow these marvellous processes? The Author of these wonders understands them. None other can do.[5]

The problem is that science tells us there is nothing especially privileged about our position. Nothing is conclusive, we are eternally in 'the middle of things'. Writers frequently illustrate or try to defeat this problem by turning it into a problem of scale. Perhaps we are the wrong size. Lewis Carroll shrinks and grows his Alice in Wonderland and Jonathan Swift does the same with his Gulliver. The size changes the perspective and the meaning. Our children's stories are full of dwarfs and giants. By being one or the other we could be different, more decisively sure who we are. If we were dwarfs the size of electrons, perhaps the unreasonable mysteries of subatomic physics, would resolve themselves. Or, if we were giants the size of galaxies, Einstein's relativity might appear to be no more than the commonest of common sense. The point is that science shows us there is nothing special about the way we happen to see things, nothing special about what the universe looks like from a human-sized perspective. In short, there is nothing special about us.

'We live', the astronomer John Barrow has written, 'in the in-between world . . . betwixt the "devil" of the quantum world and the "deep blue sea" of curved space.'[6]

'We stand', writes the physicist Freeman Dyson, echoing the thought, '. . . midway between the unpredictability of matter and the unpredictability of God.'[7]

All the maps and powers of science never seem to refer back to us. We are just expected to get on with things while this fabulous new truth unfolds.

'We feel', wrote the philosopher Ludwig Wittgenstein, 'that even when all possible scientific questions have been answered, the problems of life remain completely untouched.'[8]

This is the heart of the matter. We know science is effective and we know that it tells us that it is in pursuit of the truth of a real world. But is it the Truth? Is it our Truth? Do its awesome powers mean that science must be far more than a way of doing things? Hawking, by invoking God, says it is. He says it is a potentially conclusive way of knowing everything – that it is the Truth.

This is where the history must begin. We have seen that we are in possession of an unprecedentedly effective way of understanding and acting called science. We have seen that this way is intolerant, restless and ambitious, that it supplants religion and culture yet does not answer the needs once answered by religions and cultures. In order to understand what to do about this, we need to know more about what it is, how it came to be and how the people of our age have attempted to come to terms with its terrifying success.

2
The birth of science

What now, dear reader, shall we make of our telescope? Shall we make a Mercury's magic wand to cross the liquid aether with, and like Lucian lead a colony to the uninhabited evening star, allured by the sweetness of the place?

Kepler[1]

In 1609 Galileo Galilei looked through a telescope at the moon. It was a moment of such significance for the world that it has been compared to the birth of Christ. For, as at Bethlehem, it was a moment when the impossible entered human affairs.

Any statement of where anything as vast and vital as science began is likely to be contentious. One view would be that science did not begin at all, it has always been with us. As I have said, this idea seems to me to be pointless and wrong. The basis of this book is that science is a new element in the world that requires new responses.

Certainly it had antecedents. Some theories suggest, for example, that science sprang from the arts of medieval and Renaissance magic. Others that it was born of the revival of classical learning that swept through Europe in the fifteenth and sixteenth centuries. All such theories have their weight and explanatory power and all point to an ancestry, a family tree of the elements that formed science. But my point here is that, though it may have taken centuries for these elements to form and converge, the moment when they finally coalesced was explosive rather than evolutionary.

It was a moment lasting perhaps a decade or two. But I choose 1609 and the moment Galileo looked through his telescope because it contains all that was new and revolutionary in science.

I also choose the moment because its place in the mythology of our culture is secure and unarguable. It is a good beginning precisely because it is part of the familiar imagery of our schoolbooks. The spectacle of a man in early seventeenth-century clothes peering through a primitive telescope is an icon of our understanding of the modern world.

Yet all Galileo did was to believe the evidence of his own eyes, aided as they were by the crude optics of his telescope. And all he saw was that the lunar surface was rough-hewn and mountainous. It was, he concluded with an entirely new type of confidence, remarkably like the earth.

There was nothing romantic about the man. He was practical, stocky and redheaded, politically aware and commercially astute. He had a flare for attracting the interest of merchants in his devices and for convincing cardinals of the importance and, for a time, the orthodoxy, of his science. He was, we might say as we so often do of the sophisticated products of the Italian High Renaissance, a modern man. But, with Galileo, the cliché has an added dimension – for he invented the modern.

What is necessary, therefore, is to understand what, exactly, Galileo 'saw' and how what he saw became that which we all now see.

He saw, as I said, the impossible. It was impossible because the entire culture from which Galileo sprang was based upon the 2000-year-old certainty that the moon, like all else in the heavens, could not be like the earth. It was different because it was not made of the same substance. It was made of celestial matter which was pure and unchanging. Any imperfections seen by the naked eye were either minor flaws brought about by its proximity to earth or deficiencies in our organs of sight. The poet Dante had taken the latter view, he called the moon the eternal pearl. This was not just poetry, it was medieval science. And medieval science was not like ours, subject to change and modification. It was ultimate truth. It was the complete explanation endorsed by the authority of God and the Church and conclusively deciphered by the great human adepts from Aristotle to Saint Thomas Aquinas.

This science – better, perhaps, to call it 'wisdom' – that existed before Galileo was different in every respect from the science –

our science – that ruled after that moment in 1609. Its foundation was neither observation nor experiment, but authority understood through reason. And it was inseparable from that vast edifice of explanation, the Roman Catholic Church. Protestantism had challenged this monopoly long before Galileo, but even the most radical Protestant reformers retained the conviction that knowledge of the world could only be knowledge of a system utterly determined by and clearly demonstrative of the love and infinite wisdom of God. And, if this were so, proof of the truth of any cosmic system did not wait upon human verification. Rather it was authoritatively true. Its human documentation was merely an assertion of this authority. In a fundamental sense, neither Galileo nor any other individual was *qualified* to question any of its details.

In Dante's *Divine Comedy* (1321), the highest artistic expression of this pre-Galilean vision, the cosmos is shown to be a machine constructed around the drama of salvation. This machine was built by a tradition of thought that flowed from Aristotle through Aquinas. It was constructed by great individuals but they had become, to the orthodox mind, merely the instruments which had exposed the nature of Divine Reason. To Dante, middle-aged, besotted with love for his teenage Beatrice and, in his dreams, entrapped within the timeless mechanism of his cosmos, it provided a precise and unarguable identity. It was perfect, complete and utterly rational. The whole point of his poem is the ultimate revelation of exactly where the poet stood.

The only problem, as Galileo now knew, was that it was visibly wrong. The map on which Dante had located himself was defective.

In the universe of Catholic orthodoxy the earth was not only at the centre of the universe, its composition was also fundamentally different. The earth was at the centre because all that was changeable, all lumpish, heavy matter had, impelled by its intrinsic nature, found its way to this mid-point. We dwelled upon this brute surface, ourselves subject to change and decay, but partaking of the great narrative of salvation because of the reality of Christ. We swung, as on a pendulum, between animals and angels. The heavens, in contrast, were unchangeable and pure.

Celestial matter was refined and perfect, utterly different from that of earth.

From our own perspective it is, at first, difficult to understand the centrality of this image to the Catholic mind of the early seventeenth century. But, in an important sense, we can only understand who we are by grasping how different imaginations worked. So, if we loosen the grip of our modern education, it is possible to see the power of the idea of celestial purity. To the naked eye, innocent of modern cosmologies, the heavens *do* look clean and unchangeable. The night sky can still be a majestic image of ultimate peace and calm. In contrast, our selves and our world are messy, dirty and chaotic.

What I wish to convey is that there is a human truth to the pre-Galilean vision, even if, now, we would patronizingly relegate that truth to the realm of the 'poetic'. I do not believe truth is so easily divisible.

Beneath the modern 'scientific' gaze, however, Aristotelian cosmology and physics appears to be a quaint, improbable system that could never have survived the most elementary evidence of observation. Furthermore, it seems strange that a system devised by a pre-Christian philosopher should have such a hold on the dogmas of the Christian Church. Even the awesome authority and grandeur of the greatest mind of antiquity must surely bow before the transformation of history that occurred with the birth of Christ. After all, Virgil, the supreme poet of the classical, pre-Christian world, had been alloted no more than a place in Purgatory rather than Paradise in the *Divine Comedy*. In the mighty scheme of things, to be born before Christ meant that you were excluded from Paradise.

But the reason Aristotle's authority survived the onset of Christianity intact was the power and consistency of ancient, classical thought. This was too great simply to be sidestepped by the intellectuals who codified the Faith of Christ. They did not deny that Christianity possessed the unique Truth. But they knew that it had nothing that even the refined doctors of the Church could convincingly place alongside the splendours of Aristotelian cosmology. In addition, the medieval rediscovery of classical learning was worn as the badge of its civilization, a civilization that was emerging in all its humane sophistication and splendour from the long night of the Dark Ages. An intelligent acceptance of pre-Christian wisdom was, in fact, a way of celebrating the triumph of Christianity, its power over all knowledge.

The ultimate expression of the need to unite Christian and classical wisdom was St Thomas Aquinas's *Summa Theologiae* of 1266. This unfinished work, described by a modern theologian as the one moment in history at which the art of theology had a theory, represents the pinnacle of intellectual Christianity. Standing like a great mountain between our age and that of Christ Himself, it can now be seen as the pivot upon which the history of the Faith's struggle with itself is balanced.

The *Summa*, in effect, said everything. It was a theological version of the Theories of Everything towards which Hawking and the whole of modern physics aspire. It endorsed and refined Aristotle and established the primary synthesis of the medieval mind by uniting his thought with the revelation of Christianity. Its unity and completion as well as its confidence are qualities shared with and demonstrated by its fabulous and extravagant contemporaries: Europe's Gothic cathedrals. This was the age of medieval humanism and its triumphant attempt to sweep the whole of creation and history into a single intellectual structure. The new architecture embodied the new spirit in its ingenuity, its clarity and the cerebral passion of its structures.

Everything in these buildings was reasonable and efficient. There was no stone where there was no stress, nothing was redundant and nothing concealed. Look at a Gothic cathedral and you can *reason* your way to its truth. The humble figures of peasants in the stained glass of the great cathedrals were included in an architectural narrative that culminated in the saints and God Himself. The churches were the cry of a triumphant Christian synthesis, an emotional realization of an intellectual perfection. They were what Aquinas's *Summa* would have looked like.

But the rationalism these buildings and this book celebrated was ambiguous. As the historian Hugh Thomas has pointed out, these soaring, arched spaces were perfect for hearing music, less good for deciphering words. The cathedrals celebrated a rationalism, but subverted reason. They were dreams of unity and completion, not temples of speculation. In a fundamental sense there was nothing to speculate about. All the speculation had been done. You could work out how the cathedrals were built, but you could not see them in a new way. You could only work towards an understanding that had already been established.

Aquinas's arguments were superbly sophisticated. He came from the Dominican order and he did not share the rival, humble Franciscan view of Christianity. Rather he saw the Faith as a primarily intellectual structure, a mechanism accessible to reason. As was his personality, so was his age, a period of newly discovered confidence in the power of the human imagination. Medieval humanism would have found it impossible simply to cast aside the achievements of antiquity, so it was Aquinas's destiny, thanks to his analytical genius, to show that the classical and the Christian could be united.

It became the project of the entire era. The cathedral at Chartres was completed thirty years before the *Summa*. It is, perhaps, the most eloquent of all expressions of medieval humanism and the Gothic spirit – an overpoweringly consistent celebration of an all-inclusive intellectual synthesis. After many visits Chartres still renders me speechless with the certainty and unity of its vision. The building is obviously beautiful, but also brutal in its single-mindedness. And there, in its stonework, are carved the compact, intense figures of Pythagoras and Aristotle, the mathematician and the presiding genius of the classical world at one with the triumphant glory of Christianity.

Aquinas's goal was to provide a medieval structure subtle enough to compete with the ancients. By the end of the twelfth century the works of both Aristotle and Ptolemy, the greatest astronomer of antiquity, had been translated and were thus available to scholars throughout Europe. The majesty of their schemes was inescapable and the response of the Christian intellectual, bereft of competitive cosmologies or physics, could only be to accept and unify the ancient with the modern.

But this unification carried with it the dangers that all modernizations bring to all faiths at all times – the dangers of dilution and compromise. Aquinas's summation, for example, accepted that parts of the Bible were not literally true. Some stories were acknowledged to be illustrative metaphors designed to help simple minds grasp the underlying reality. This had happened before – Aquinas probably derived the idea from St Augustine – but its acceptance in the *Summa* is crucial. It seemed reasonable enough, an acceptable 'sophisticated' kind of analysis. But it was a tiny rent in the fabric of Christian belief that was never again to be closed and, in the nineteenth century, was to spread and rip apart the entire material foundation of the Faith.

The further danger was that Aquinas represented a profound modification of the original Christian impulse, simply because he was such an intellectual master. As Max Weber, the great sociologist of religion, pointed out, 'a considerable portion of the inner history of the early Church, including the formulation of dogma, represented the struggle of Christianity against intellectualism in all its forms.'[2] This was the style of the Franciscans. The faith had glorified the poor in spirit rather than the scholars; indeed, great subtlety and learning have always been felt by a certain type of Christian to be especially fertile ground for sin. But the Thomist – the label given to Aquinas's thought – synthesis was intellectual in the extreme and patrician in tone. In detail it was to be contradicted by everything that was said by Galileo and Newton. But, in the important sense that it glorifies the power of human reason, it has to be understood as a kind of stylistic foundation of modern science. As such, of course, it is possible to see Aquinas as one of the greatest enemies of the old faith even as he was to become the creator of the new.

But what was this classical–Thomist universe that ruled in the minds of men at the moment Galileo looked through his telescope? Aquinas employed it as the physics of his system to balance the metaphysics and theology of Christianity. The rational authority of Aristotle was combined with the moral and human force of Christ.

This old cosmology was, first of all, very powerful. In the hands of Ptolemy, classical astronomy had become a means of calculating and predicting the movements of the stars and planets with extraordinary accuracy. It was, apparently, shown to be true every day by the precision with which the heavens obeyed the Ptolemaic model.

But it was Aristotle who dominated Thomist thought. He was, and remains, the very emblem of human wisdom. Born in 384 BC, he was, for a time, the tutor of the young Alexander the

Great. His influence was centred upon the school he founded in Athens. But, after the death of Alexander, he was charged with impiety. Unlike his predecessor, Socrates, he did not obligingly kill himself. He fled and died a year later in 322 BC. His style was dispassionate, cool and analytical and his intelligence universal. Such details are important in his case because they emphasize the earthbound, essentially 'sensible' nature of his thought – especially in the fact that he chose flight rather than the traditionally noble option of suicide. Aristotle's physics and cosmology, that were to be the basis of medieval thought, were, above all, reasonable, however strange and 'poetic' they might now seem to us.

His universe was based upon the division of creation into two realms: the sublunary and the superlunary, meaning simply below and above the moon. The moon marked the crucial dividing line. Beneath it was all that was changeable and subject to decay and all that was constructed of the four elements – earth, air, fire and water; above were the changeless, perfect heavens. The moon was at the point of transition, so imperfections that were observable on its surface may, in some interpretations, be due to its proximity to earth. But that did not mean that one could say, with Galileo, that the moon was *like* the earth. Even if corrupted by proximity, it was still made of celestial matter.

The superlunary realm was conceived by Aristotle as a series of nesting, concentric shells whose rotations explained the movements of the celestial bodies – the stars and planets. The shells were driven by a kind of friction drive that was transmitted downwards from one to the next. On the outside was the *Primum Mobile*, the shell whose movement drove all the others.

The shells were made of a crystalline solid known as the aether and Aristotle concluded there were fifty-five of them. The drive mechanism applied throughout the system and its implication that the whole universe was directly interactive provided the foundations of astrology. Events on earth could be seen to be mechanically linked by the friction drive to events in the heavens. This view was to bind together astrology and astronomy for almost 2,000 years after Aristotle and the imaginative attractiveness of the idea has ensured that astrology is still with us today. Astrology has always, however, dwelled uncomfortably with 'official' Aristotelianism – in the Middle Ages the Church attempted to suppress the study of astrology as inconsistent with the Christian doctrine of free will. This indicates the type of tension involved in unifying Christianity and classicism – a tension between the logic of physics and of theology.

Ptolemy in his *Almagest* in about AD 150 provided an infinitely more complex system which explained, through its system of deferents and epicycles, every perceived motion in the heavens. For a system which we now consider to be entirely 'wrong', it was spectacularly accurate. Ptolemy, for example, calculated the distance of the moon from the earth as 29.5 times the earth's diameter. Our figure is 30.2.

What is significantly unclear in Ptolemy is whether his system is an attempt to describe what the universe is like or whether it is a model which provides an explanatory system for the movements of the stars and planets. This ambiguity will surface again and again in the history of Western thought. Even in our own century atomic theorists have been divided as to whether our image of the atom is a convenient fiction or a picture of reality. The point is crucial to our understanding of what science might be, of what its 'truth' might consist.

Whatever Ptolemy meant, his system represented another side of classicism. It was a side that was closer in spirit to the classicism of the Renaissance and of modern science than to the classicism of Aquinas and the scholastic theologians, for it leant more heavily on the evidence of observation. But it remained a pre-scientific theory, in my sense, because it lacked the particular combination of reason and observation that was finally to occur in Galileo.

What all pre-Copernican and pre-Galilean systems shared was an overwhelming and radical difference from anything that we believe today. Our puzzlement at the medieval need to believe in the authority of the Aristotelian system betrays the truth that we are very far indeed from the imaginations of the scholastics and the faithful. Simply describing their systems cannot encapsulate the sheer difference of classical wisdom from our own. To understand the present we need to grasp the way imaginations were formed by medieval cosmology because that will show us how different is the way our science has formed us.

There are many ways of defining this difference but all can be best expressed by an understanding of the Aristotelian notion of causality. This shows how the simplest of our daily perceptions can actually be changed by an intellectual and imaginative climate.

Our way of understanding causality is a condition of our understanding of the world. It is simply a way of answering any question about why things happen. In everyday life it is a trivial matter: a ball moves because it is kicked, a stone falls because it is dropped, we cry because we are unhappy. The problem about defining causality is that even these elementary examples can be infinitely extended into realms of kinetics, gravity, psychology and on until we attain the furthest reaches of the large scale universe or the small scale particle as the founding cause of why we may have shed a tear.

So it is a simple notion, but it seems essential to our perceptions and it leads rapidly to complexity. Such extrapolations into chaotic multiplicity make the precise nature of causality difficult to establish and Aristotle provided a typically elaborate solution.

For him there were four types of cause just as there were four elements. The types were: material, efficient, formal and final. So the material cause of a house may be its bricks and mortar, its efficient cause would be the act of building it, its formal cause would be the design conceived by the builder and the final cause would be the creation of an object in which one could live.

Note the essential benevolence of this system as well as how alien it is to our modes of thought. The final cause gives a beneficial conclusion to the whole activity, and is implicit at the start of the process. Such a view makes sense of the cosmos. The great mechanism of nesting shells can be seen to exist solely for the purpose of the earth, the tiny object at its centre, the final cause of the whole system. There was nothing odd to Aristotle or Aquinas, as there is to us, about the idea that the earth must be at the privileged centre of all creation.

However, we may find the idea of a final cause almost acceptable in the case of a house. But consider a plant or a mountain. For this fourfold causality applied to the entire universe and it assumes what we now usually consider to be an impossibility: the simultaneous existence of cause and effect. If we study the evolution of an animal from its ancestors we would find it absurd to think that, somehow, man was already present in an ape or an amoeba, that we existed fully formed within such organisms.

In practice we are now only really concerned with efficient causes and we tend to reduce all the other categories into different variations of the idea of efficient causality. The cause of man evolving from an ape is natural selection – the biological equivalent of building the house; it is not some innate tendency within nature to produce men.

The great difference from our own way of thinking is that Aristotelian causality included people. The human race was the point, the heart, the final cause of the whole system. It placed our selves definitively upon the universal map. Our system does the opposite. It presents us as accidents. We are caused by the cosmos, but we are not the cause of it. Modern man is not finally anything, he has no role in creation.

The concept of an 'innate tendency' is just as important if we are to understand the overthrow of Aristotelian physics. His causal system entailed a view that objects possessed within themselves some predisposition towards their characteristic behaviour. A stone fell because, being brute matter as opposed to celestial substance, it was drawn towards the centre of the earth – the centre of the universe. That was the nature of a stone, that was what it meant to be a stone.

Two further details of ancient physics are worth mentioning for the way they were to be rejected by the new science. The Aristotelian concept of space made it clear that a vacuum was impossible. Space, being only definable as where an event occurred or an object rested, could not be evacuated. Matter and space were inseparable. Finally, Aristotelian dynamics pursued its own alien logic: a stone flying through the air, for example was being propelled by the turbulent air in its wake.

In almost every detail the Aristotelian system contradicts our own. Again, however, I must stress that we can grasp its appeal. As I said, our 'innocent' response to the heavens might easily confirm a classical cosmology. Similarly there is nothing that is *obviously* wrong with Aristotle's causality or his idea of innate tendencies. If we turn off our own logic for a moment, we can see this other logic with perfect clarity. We have been 'taught' our way of seeing as surely as Medieval Man was taught his.

But all of this was, in the Middle Ages, the only powerful, complete and coherent explanation to hand and it was conclusively endorsed by authority. It was the Truth. Through Aquinas's unifying efforts, Aristotle's words became more than speculations or even explanations; they became dependent on and supportive of the authority of the Church. To disbelieve or attempt to disprove Aristotle was thus to mount a challenge to the Faith. Like the cathedrals this was not a structure of speculation open to all men to debate. It was a single, authoritative solidity, constructed of patrician, scholarly dreams.

There were sporadic challenges, of course, but it was not until the sixteenth and seventeenth centuries that the dissident momentum became overpowering. Separating all the strands that led to this revolution, including, as they do, the Renaissance, the Reformation, the Counter-Reformation, the discovery of America and certain crucial technological advances – notably the invention of the telescope – would be a gigantic task. But a central tendency is clear: the world and the universe were found not to fit the Aristotelian/Thomist model and that sudden discontinuity began the long, painful process of the undermining of dogmatic authority.

The story, as I said, can be picked up at almost any point. Modern science might be said to have begun with Aquinas in that his synthesis was the theory which, by its own destruction and by its radical modification of early Christianity, created its successors. Religiously there was the individualizing, existential anguish of Martin Luther and the whole, vast drama of the Reformation. There was the sudden flowering of scientific curiosity and knowledge in the sixteenth century. There was the gigantic efflorescence of creative genius we call the Renaissance. This was also a return to classicism, but not in the medieval sense. For this was classical humanism, a belief that man may be the measure of all things. The architecture of the Renaissance was classical, not Gothic. And in a classical church the spoken word can be heard. Reason rules and reason can be lethally curious.

But perhaps the most vivid precursor of the new world that

was about to be born is also the simplest and the most celebrated – the discovery of America by Christopher Columbus in 1492.

Ptolemy had said the earth was spherical and Columbus had picked up this improbable idea as a young man. In the confident, mercantile, expansive mood of late fifteenth-century Spain and Italy, the exploitation of this hypothesis to establish an alternative route to the Indies was an obvious, if extraordinarily courageous, project. One can see the revolutionary momentum even in the simple fact of his decision to sail. A hypothesis was to be judged against the real world. There were clear commercial reasons for such a test of our knowledge and commerce has always been willing to modify doctrine in the name of profit. But the real significance is that authority was being tested to see if it was right.

As it happened, authority was found to be both right and wrong. The earth was round, but, westward, America, previously unknown to every authority, lay between Europe and the Indies. This discovery was not at once a threat to the prevailing order of orthodoxy. Europe was shaken by the idea of this undiscovered wilderness, but the new continent could be incorporated into existing world views. It did, however, create a terrible open-endedness. If an immense continent could lie beyond our immediate perceptions and knowledge, what else might there be to find?

The lesson was learned at once. The motto on the crest of Ferdinand and Isabella – the Spanish monarchs who had financed Columbus – had been 'ne plus ultra' – no more beyond. This motto was heraldically entwined about the Pillars of Hercules, the legendary edifice taken to represent the limits of the known world. After Columbus had discovered America the 'ne' was removed. The motto now read: beyond this more. As a royal gesture this was not modesty, it was the proud statement that the Spanish monarchs were the gatekeepers of the New World. But, as philosophy, it was much more. 'Plus ultra' was, more than a century later, to be adopted as the slogan of the new science. It celebrated the possibility of infinite progress, of unlimited possibilities of knowledge, of an eternal voyage.

'Let this effect of nature,' Pascal was to write, 'which previously seemed to you impossible, make you know that there may be others of which you are still ignorant. Do not draw this conclusion from your experiment, that there remains nothing for you to know; but rather that there remains an infinity for you to know.'³

The revelation of America introduced the possibility of radical ignorance into the human mind, the possibility that there was an infinity for us to know. It exposed the fatal incompleteness of all previous models of the world. The refined, subtle, fabulously complex systems of Aristotle, Ptolemy and Aquinas assumed completion. They drew boundaries. Ignorance could only be of the details.

But, with Columbus, came the revelation that we might, in reality, know scarcely anything at all. This continent had, after all, remained hidden from our gaze, however reasonable and authoritative we might have believed ourselves to be. The implication of this was that knowledge was a dynamic state of affairs. It was progressive, a process of annexing the unknown. And it was this movement from the static to the dynamic that was to characterize the overthrow of the systems of antiquity and replace them with the dominant, dangerous, restless modern form of knowledge that we now simply call science, but the intellectuals of the seventeenth century knew as *Scienza Nuova*, a term apparently coined by the sixteenth-century mathematician Niccolo Tartaglia.

'*Nuova*' – new – was the point. Galileo was always to insist on the absolute novelty of his own discoveries and, from the seventeenth century to our own day, our culture was to embrace novelty with accelerating intensity and with formidable consequences. The need for innovation was born with science.

The idea of Columbus's project and his achievement prepared the European mind for what was to come next. In the year 1543 there were two further precursors. Two books were published. One was by the Fleming Andreas Vesalius. It was called *On Fabric of the Human Body* and studied man as an empirically observed anatomical phenomenon. It was a return to the Greek medical tradition which had reached its climax with Galen. But its significance here was that it elevated a humane focus on the things of this world *as things* rather than as symbolic representations of some other reality. This was the cold, clear-eyed classicism of the Renaissance, not the visionary, dreaming unity of the Middle Ages.

But it was the other book that changed the world, that literally shifted the ground under the feet of mankind. Written by the Pole Nicholas Copernicus this was *De revolutionibus orbium coelestium* (on the revolutions of the celestial spheres). Nothing about the man – a conservative disciple of Ptolemy and an adherent of Aristotelian physics – or the book – a highly technical and mathematical treatise – forewarned of the revolution they would launch, that he was playing John the Baptist to Galileo's Christ. Indeed, as with Ptolemy's great work, it was perfectly possible to take the book either as a tool or a description. As a churchman this was what saved Copernicus from the persecution that would later be imposed on his followers. The authorities were free to interpret *De revolutionibus* as simply a useful model that would 'save the appearances' – that is, one that would work as a predictive and descriptive mechanism, but would have no particular truth-value attached.

The problem with the 'saving the appearances' excuse was the same problem that was to confront all who attempted to question the triumphant march of science over the next 450 years. This was the problem of success, of devastating effectiveness. The Copernican system replaced geostasis – a stationary earth – with heliostasis – a stationary sun – and geocentrism – a central earth – with heliocentrism – a central sun. In doing so it created a system of surpassing clarity and economy compared with everything that had gone before. Copernicus could compute distances and times that would previously have seemed far beyond the reach of human knowledge, far beyond even Ptolemy. In fairness it should be said that it took Johannes Kepler to make the system work fully. But, either way, the one overriding message of Copernicanism was: it worked. One could still argue that it was a successful model and no more, but the argument was always weakened by the overpowering effectiveness of the system.

This was the beginning of the process I described in my first chapter. The new way of understanding begins to appear in the world as penicillin might in a remote, primitive tribe, and its effectiveness is such that it cannot be removed. Tribal gods are humiliated by this new, powerful truth and they become museum pieces, drained of their old meanings and authority. This was exactly what was about to happen to Aristotle and Aquinas.

The point was that it made far more sense to believe that Copernicus had stumbled on an idea that equated with the truth much more closely than any of the dreams of his predecessors than to believe he had simply come up with a better model. Doubt had begun to threaten the old gods.

Even the Protestant radicals saw the scale of the challenge represented by heliocentrism. Luther called Copernicus an 'upstart astrologer' and Calvin asked: 'Who will venture to place the authority of Copernicus above that of the Holy Spirit?' The answer was to be Galileo.

The heliocentric universe was a fundamental challenge to the orthodox view, even if it was initially neutralized by the ambiguity of the claims of the Copernican system. A central sun removed the earth from the focal point of creation. It could no longer be a privileged place, rather it was just one more planet among others. Yet the whole drama of Christian creation as previously understood depended on the earth and mankind being the reason for it all, the Aristotelian final cause. We were the end, the purpose, the rational axle about which the great aetherian shells rotated. Pull the Copernican thread and first Aristotelian/Thomist physics would unravel and then the faith itself.

With Luther three years from the end of his anguished life and the Reformation in full flood, the Catholic Church needed at least to keep the intellectual fabric intact. It was, therefore, far more pressingly serious that any conceivable imaginative crisis. It was politics.

But, over the next 150 years, the thread was to be pulled without mercy. *Scienza Nuova* turned out to have an energy and vital power hitherto unseen in the history of human ideas. Some mysterious force seemed to have been unleashed by the human mind which could not be prevented from breaking through all previous boundaries. It was irresistible because it worked.

And this force was served by genius – most spectacularly in its early life by Galileo and Isaac Newton. Successively these men and their followers found that ever more crucial elements of the observed world did not agree with the demands of authority. Even more importantly they found that the human mind could know more – far more – than had ever before seemed possible. Alongside the despair of the modern, there is also the arrogance. With our displacement from the centre of the cosmos, we began to become tragic gods, defeated by our destiny and yet more powerful than anything else in creation.

One other key element of this revolution needs to be recorded. This was the revitalization of science's own language – mathematics – by the introduction of Arabic numerals into Europe. In fact, the system derived originally from India, passed through the Arab world and into Spain. There is a strange kind of sad irony at work here. It was the Muslim occupation of Southern Spain that caused the introduction of modern numerals, complete with the invaluable innovation of the number zero, into Europe. Yet it was Ferdinand and Isabella, the patrons of the New World and of new knowledge, who finally drove the Arabs out of Spain. Boabdil, the last Arab King of Spain, was evicted from his capital of Granada in 1492, the year of Columbus's voyage. Arab genius had devised the primary language of the new science, but the decline of their power was to leave the Arabs as only second-hand beneficiaries of the revolution for which they had written the text. When Boabdil looked back and wept for his lost Alhambra, he was weeping for far more than he could ever know. And the modern Arab leaders who confront the West – Gadaffi, Saddam Hussein – are his descendants in loss.

Effective mathematics was the language in which the new science found it could speak to itself. Once numbers had taken on their modern form, they became instruments of almost bewildering power. Through some obscure magic these human creations seemed to be able to gauge the entire universe. As with science itself, this extraordinary power made them magical, effective beyond the dreams of reason. I shall examine later the complex issue of whether they can also be the gauge and model of the mind of man – for that is what many modern scientists now believe.

As well as maths, science also needed the age's rapidly developing technological skill. This is a mild paradox. Technology is logically the application of science, so it must come second. But the craftsmen of the Middle Ages and the Renaissance devised technologies that, in fact, preceded and inspired the science that was to realize their full significance.

The most important developments are obvious. They are the ones that endowed man with control over space and time. Imagine, for a moment, a world without clocks, maps or telescopes. This was the world before the fourteenth century. Time was an imprecise concept, place was uncontrollable in the sense that it was perceived as not existing if nobody had been there. Finally, the great distances of the heavens were subjected to the limitations of the human eye.

Technology began to change all this when, in the early fourteenth century, the counterpoise clock was invented in Germany and public timepieces at once began to spread throughout Europe, superseding the old systems of hourglasses and sundials. Time became a precise notion. One could live one's life with perfect accuracy. The placing of the clocks in central, public places emphasized that time now had an impersonal authority, an existence beyond ourselves and yet one which we could now control through our knowledge. Subjective time – our own private sensation of duration – was implicitly humbled and our modern obedience to objective, measured time was born.

Great sea-voyages like that of Columbus also inspired the compilation of maps. In the sixteenth century Mercator, the Latinized name by which the Flemish geographer Gerhard Kremer has come to be known, devised his system of projection which allowed navigators to plot correct compass bearings. A map, as I suggested in my first chapter, is another potent form of the objectification of our knowledge. Without one we may simply say we do not know what lies beyond the next hill. With one we can contain our ignorance with the tools of a projection and a grid. We may still not know what lies beyond the next hill, but we can delineate the margins of our ignorance.

Clocks fixed time and maps contained space, both throwing nets over the unknown and both establishing absolutes beyond ourselves. Space and time in maps and clocks existed without us and beyond us. In understanding the present it is worth noting that the full impact of these technological developments was not felt until our own century. The benefits and disciplines of maps and clocks began as the privilege of a wealthy elite, but did not directly affect the life of the majority. Industrial development changed all that by making technology central to all our lives. The climax of this process came in the twentieth century when knowing the time and our place became the foundation of every life.

But it was in the seventeenth century that this foundation was laid. Pre-eminently in the thought of Isaac Newton time and space were to become the unalterable conditions of the cosmos – absolute space and absolute time.

The implications of all such projects, such possibilities, slowly permeated the European mind as if in preparation for Newton's birth. But, initially at least, no technological innovation could match the telescope as a means of shocking, subverting and ultimately transforming the entire system. Tales of its invention vary. Some say there were telescopes around in the Middle Ages, some that it was first made in Italy, some in Holland. But invention was only half the story. The other half was ensuring that a telescope found its way into the hands of a genius and, in 1609, one did.

'A report reached my ears,' wrote Galileo in that crucial year of 1609, 'that a certain Fleming had constructed a spyglass by means of which visible objects, though very distant from the eye of the observer, were distinctly seen as if nearby.'[4]

Galileo's great contemporary, the astronomer and mathematician Johannes Kepler, was less restrained: 'What now, dear reader, shall we make of our telescope? Shall we make a Mercury's magic wand to cross the liquid aether with, and like Lucian, lead a colony to the uninhabited evening star, allured by the sweetness of the place? Or shall we make it Cupid's arrow, which, entering by our eyes, has pierced our inmost mind and fired us with a love of Venus? . . . O telescope, instrument of such knowledge, more precious than any sceptre! Is not he who holds thee in his hand made king and lord of all the works of God?'

The inflamed rhetoric captures all the dangers of the enterprise. A telescope could allow a man to challenge God. Its use was a Promethean crime, an act of pride. Indeed, some orthodox believers in the seventeenth century compared the practitioners of the *Scienza Nuova* with the builders of the Tower of Babel – they too would be struck down into chaos for their presumption.

Of all the complex convergences that produced the new science, the one that catches our imagination most is a coincidence. For Galileo and one of the earliest telescopes to be in the world at the same time looks now like a coincidence of awesome beauty and symmetry. Into the hands of one of the greatest of all astronomers and physicists had fallen this Promethean instrument that would allow him to look into the heavens and *see* the new order. The word 'see' embodies the revolution. With his telescope Galileo did not have to construct elaborate models to accord with the crude simplifications imposed by the naked eye. He did not have to play games with 'saving the appearances' to pacify the theologians. He could test his ideas in a way that was unavailable to either Ptolemy or Copernicus. Like Columbus setting out to test if the earth was round, Galileo could test the entire fabric of the cosmos. Like Columbus he could span great distances, he could extend the reach of man. He could forever voyage.

Or perhaps I mean he could extend the reach of *a* man. For this new knowledge was being discovered by individuals working alone. Authority was to be subjected to a single man, an individual, putting his eye to a simple man-made object.

'In science,' wrote Galileo with his usual devastating clarity, 'the authority embodied in the opinion of thousands is not worth a spark of reason in one man.'[5]

The statement is shocking, heretical and impossible. But Galileo had this new confidence, this new wisdom that sprang from his eye, the telescope and his reason. The fabulous dreams of Aristotle, Aquinas and Dante were about to be revealed as just that . . . dreams. We were about to wake up, or so we now think.

Even without a telescope, Galileo had already severely damaged the Aristotelian universe. In 1604 he had proved that a new star in the constellation Serpentarius was a genuine star – in other words it was located in the superlunary spaces. Since it had just appeared this demonstrated that change was possible in the heavens. This was a timely and, for Aristotelians, an ominous repeat of the observation of 1572 by the Danish astronomer

Tycho Brahe of a nova in Cassiopeia. Both could be shown to have no measurable parallax. In other words they did not appear differently from different points on the earth's surface. If they were close, they would do. They must, therefore, be very distant from the earth – in the supposedly unchangeable heavens.

But it was the telescope that sealed Aristotle's fate, and ours. First Galileo saw the moon. He saw and even calculated the height of its mountains with an accuracy that is endorsed by today's measurements. He demonstrated, by observing a secondary level of illumination on the surface of the moon, that the earth shone, *like the other planets*. He discovered the moons of Jupiter, countless new stars . . . in fact, Galileo discovered what we call the universe.

It was a potential catastrophe, but, for the moment, he and *Scienza Nuova* were safe. There was a temporary lull in the defensive fervour surrounding the old wisdom. These were the liberal-minded years of the late Renaissance in Italy and such discoveries were initially greeted with excitement rather than identified as threats. Indeed, Cardinal Maffeo Barberini wrote a poem in praise of Galileo. The irony was that Barberini was later to become Urban VIII, the Pope who ordered Galileo's trial by the Inquisition on charges of heresy.

It should be said at this point that the ensuing battle between orthodoxy and science is not the simple opposition it might appear to be. Science, the 'spark of reason', from the beginning, was not merely seen as a threat to the edifice of dogmatic religion. It was also identified as the enemy of the new humanism of the Renaissance. The poet Petrarch had wondered what the final good could be of the explosive growth in the knowledge of nature that he saw around him, even before this new knowledge became definable as science.

'Even if all these things were true,' he wrote, 'they help in no way towards a happy life, for what does it advantage us to be familiar with the nature of animals, birds, fishes and reptiles, while we are ignorant of the race of man to which we belong, and do not care whence we came or whither we go.'[6]

The discontinuity between scientific knowledge and human happiness echoes across the ages. When Albert Einstein's wife was asked if she understood relativity, she replied: 'Oh, no, although he has explained it to me many times – but it is not necessary to my happiness."[7] We might ask: for what, then, *is* it necessary?

So there was something clairvoyant about Petrarch's melancholy. He saw science in its pessimistic, inhumane aspect. It was a distinctively human creation and yet, perversely, it had no place for humans and their cares. Indeed, its very success came to seem based on the exclusion of the merely human. At one level this is the beginning of all subsequent humanist suspicion of science; at another it identifies a profound contradiction in the nature of our knowledge to which I shall return.

But, in the liberal climate of Renaissance Catholicism, science was tolerated, indulged and even admired as a fashionable pursuit. This honeymoon did not last long. The climate was rapidly transformed by the sudden realization, primarily among Jesuit intellectuals, of what was at stake. They saw how far Galileo had pulled the thread.

This was not, as is popularly imagined and usually taught, simply his insistence that the sun was at the centre. On this issue there was always the Copernican get-out clause of 'saving the appearances'. The real problem with Galileo was that his genius had a universal quality. He was not engaged merely in arcane plottings of the stars and planets, he was also conducting dangerous experiments with gravity and kinetics as well as speculating about the nature of matter. The first theory of relativity came from Galileo. And it was from his physics rather than his astronomy, as the writer Pietro Redondi has shown, that the most precise challenge to authority emerged.

As I have said, a central part of the genius of Aquinas had been his creation and defence of the dogmas of the faith by the adaptation of Aristotelian physics. The dogma that was to prove the most decisive and contentious was transubstantiation. This is the mechanism whereby the 'real presence' of Christ in the bread and wine of the Eucharist is explained. Clearly, even to the faithful, the bread still appeared to the senses as bread, not flesh, and the wine as wine, not blood, after the priest had pronounced them transformed into the body and blood of Christ. But the dogma depended on the conviction that they were no longer merely bread and wine. The Thomist explanation was a subtle distinction between substance and the qualities attached to that substance. The qualities – colour, taste, smell, texture – were earthly accidents, the substance was the underlying reality which, in the communion service, was changed. Hence the word: transubstantiation.

The Thomist explanation was adequate until the disputes of the Reformation began to place extraordinary pressure on what precisely was being said. The Protestants rejected the doctrine and the Jesuits, the fiercest, purest and most zealous defenders of orthodoxy, hardened their position. In the Counter-Reformatory zeal of the Council of Trent in 1545 the resistance to Protestantism became an affirmation of pre-Reformation Catholicism and its dogmas. In particular, transubstantiation was made a central article of the faith.

This left the Church dangerously dependent on a theory of physics. Evidently, if the Catholics were simply saying that Christ was present *in spirit*, then there could be no problem. The act could be understood as a form of symbolism. If they were saying he was present *in substance*, they were insisting on standing on ground soon to be invaded and conquered by *Scienza Nuova*.

Symbolically this strange dispute can also now be seen as the background of the transition from one hierarchy to another. Theology, the Queen of the Sciences, was to relinquish her crown to physics. She still reigns.

But the Jesuits increasingly felt they had to fight. The liberal leanings of Pope Urban could not be tolerated. This was the age of the Thirty Years' War and the Protestant Swedish King Gustavus Adolphus was rampaging southward through Europe. Urban was obliged to acquiesce in the defensive plots of the Jesuits. Galileo was put on trial, imprisoned and silenced.

Seldom in history can there have been such an abyss between victory in a battle and defeat in a war – four years after the trial of Galileo René Descartes published his *Discours de la Méthode* and thirty-three years later, in Cambridge, Isaac Newton was grinding his own lenses, the better to understand the nature of light.

Newton was a lens himself, the focusing glass of all that had gone before. For, somehow, he saw what it all meant, he concentrated the beams of light until they burned all that they touched.

He remains the greatest of all scientific figures because he extended the reach of the new knowledge to infinity. In doing so, he imposed on future generations the task of redefining all human knowledge. But his greatness also arises from the way he was never *simply* a scientist. His mind worked on the borderline between all forms of human wisdom: magic and science, alchemy and physics, mathematics and God. Our age is diminished by the fact that we chose to accept only one part of Newton's legacy, the part we now call science but which he called 'philosophy'.

Galileo observed, analysed and even saw the cultural crisis he had precipitated with fabulous clarity. But Newton actually imagined and described a universal system of such rigour and perfection that, even today, his accuracy remains sufficient to land a man on the moon and we have only discovered shortcomings in his mathematical generalizations in the most extreme circumstances.

Science could have stumbled on, making its painful inroads into the religiously conceived universe. But Newton seemed to end the whole process. Newtonian mechanics is not a partial explanation of the way things work; it is complete . . . as complete as Aquinas.

Understanding scientists is usually a straightforward problem of understanding their ideas. But understanding Newton is impossible without a grasp of his personality. And that personality demanded completion of thought and coherence of the world as psychological necessities.

'To force everything in heaven and earth,' his biographer Frank E. Manuel has written, 'into one rigid, tight frame from which the most minuscule detail would not be allowed to escape free and random was an underlying need of this anxiety-ridden man.'[8]

Like a metaphysician rather than a scientist, he seemed to be able to understand nothing unless he understood everything. His undergraduate notebooks were full of almost childish questions: what is light, what is fire, what is motion, what is matter, what is the soul, what is God? The one underlying question is the question of his age: what is it? Give me description, explanation, certainty.

His own destiny possessed him. He was so completely alone

with his genius that he realized that the synthesis towards which he was working depended entirely on himself down to the physical demands of his experiments. The fact that he ground his own lenses was historically significant. Unlike the adepts of the old science, he would not merely reason his way to the truth, he would physically manipulate the world into revealing its nature.

Similarly his own body would endure Christ-like suffering in the name of the search. In this extraordinary passage from a letter to the philosopher John Locke, he describes the effect of his experimental attempts to stare directly at the sun:

'In a few hours I had brought my eyes to such a pass that I could look upon no bright object with either eye but I saw the sun before me, so that I durst neither write nor read but to recover the use of my eyes shut my self up in my chamber made dark for three days together & used all means to divert my imagination from the Sun. For if I thought upon him I presently saw his picture though I was in the dark. But by keeping in the dark & imploying my mind about other things I began in three or four days to have some use of my eyes again & by forbearing a few days longer to look upon bright objects recovered them pretty well, tho not so well but that for some months after the spectrum of the sun began to return as often as I began to meditate opon the phaenomenon, even tho I lay in bed at midnight with my curtains drawn.[9]

What is striking about the passage is the way the true enormity of what Newton was doing breaks through to destroy the clinical description of symptoms. The sun had done more than damage his eyes, it had possessed him. He recovers not simply by resting his eyes, but also by fighting to eliminate the sun from his mind.

Observation and theory had provided the beginning of the descriptive control of the universe, but, at a stroke, Newton's arrogant passion seemed to provide the how and why. His gravitational constant and his laws of motion penetrated the cosmos and his absolute time and space cast an infinitely greater map than Mercator's across the heavens. This was the Theory of Everything that superseded Aquinas.

The elliptical orbits of the planets, for example, that had arisen from the Copernican-Keplerian system were explained by the combined forces of gravity and momentum. The tendency, defined by Newton, for bodies to move at constant velocity in a straight line combined with the force of gravity, measured by Newton, to place the planets in a permanent state of inconclusive falling. To derive gravity from Kepler was the masterstroke. From that all else would follow . . . *we* would follow.

The details of Newton's system and the mathematics he invented to describe it constitute one of the greatest achievements of the human imagination. The Newtonian model has been surpassed by nothing before or since, even though it may have been fundamentally corrected. But its importance, for my purposes, lies not in its exact laws and calculations, but rather in its general nature and imaginative power.

Newton was a man, as earthbound as the rest of us – he had never seen a body in uniform motion subject to no external forces, the ideal object he used to specify his laws. To define science merely as the synthesis of observations falls hopelessly short in his case. Speculation leapt far ahead of experiment and observation, precisely as it was to do again, 250 years later, when Albert Einstein proposed the General Theory of Relativity supported by the smallest possible fragments of observational evidence – a slight perturbation in the orbit of Mercury. In Einstein's case the theory had to wait for further confirmation from the deflection of starlight round the sun. Similarly Newton was to be triumphantly vindicated long after his death in 1758. Edmund Halley had used Newtonian mathematics to predict the period of a comet – 75 and a half years – he was precisely right and an awestruck world saw that the human mind had encompassed the future by explaining the heavens. It worked.

Beneath the revolutionary mathematics, simplicity was the key. Both Galileo and Newton worked on the basis that there was an underlying efficiency in nature. Explanatory systems must be the simplest and most efficient possible. In fact, this might be said to be the one conviction they shared with Aristotle; he, too, had insisted that 'nature does nothing in vain, nothing superfluous'. It was an article of faith that unnecessary elaboration could not be a part of physics. Experiment and theory involved the creation of ideals – frictionless motion or gravitational descent unhindered by air pressure – these would then be synthesized into laws which gave the most general and simplest statement of what had been observed. The laws were assumed to operate universally and local perturbations – such as air pressure or friction – were later fed in to explain precise conditions.

But Newton realized that Aristotelian physics was, in fact, far from simple. Indeed, it was incapable of real simplicity. With its emphasis on the innate properties of matter, it created a world that was a complex collection of discrete objects. Such a world was irreducible, it could not be simplified into universal laws.

For Newton knowledge could only be found in what is sometimes called 'algorithmic compression'. Say we are presented with a row of numbers. If this line is utterly random, there will be no shorter way of specifying these particular numbers in this order other than by repeating the whole row. If, however, they are not random, we can work out the principle of their succession and that principle alone will be enough for somebody else to derive the same list. The principle may be 'Add one to the previous number' or something more complicated. This principle is an 'algorithm', a word derived from the name of the ninth-century Persian mathematician Abu Ja'far Muhammad ibn Musa al-Khowarizm. If we can completely specify anything by something shorter, then it is said to be algorithmically compressible. All science is about such compressions. When we say $E = mc^2$ or falling bodies accelerate at 32 feet per second^2, we are compressing the facts of the world so as to be able to make forecasts about future behaviour.

Clearly an Aristotelian world of discrete objects each possessed of unique innate characteristics is not compressible. In Newtonian terms, therefore, nothing can be said about it. It can only be specified by complete duplication and that is not science, it is a mere catalogue. Compressing the catalogue of the world is one way of expressing the nature of the entire scientific project.

Both the simplicity and the universality that Newton demanded and achieved contained within them revolutionary messages. Nature was simple because it was neutral. Planets were not pursuing their courses because of a transcendent moral order, they were simply finding states of mechanical equilibrium, just as water seeks out its predestined level. Furthermore, any such mechanical system must be universal since this neutrality meant no part of the universe could be privileged or exceptional. Everything, everywhere was subject to the same laws. There was no pure matter above the moon and nothing particularly impure below.

Einstein was later to summarize – with typical optimism – these messages in an introduction to a book by his great contemporary Max Planck:

'In every important advance the physicist finds that the fundamental laws are simplified more and more as experimental research advances. He is astonished to notice how sublime order emerges from what appeared to be chaos. And this cannot be traced back to the workings of his own mind but is due to a quality that is inherent in the world of perception. Leibniz well expressed this quality by calling it a pre-established harmony.'[10]

Simplicity, a universal order and the objective reality of nature – the words are a deliberate bridge built to join Einstein to Newton and to define science as a continuous process of movement forward into the unknown, into the pre-established harmony. But Einstein was also defending the ideal of classical physics against new and, to him, misguided developments in the twentieth century.

Sadly for Einstein, modern science refused to conform to the ideal and, strangely, neither did Newton. For perhaps the greatest paradox of this man lies in the fact that, though he created the modern universe – neutral, mechanical and devoid of value – another side of his personality reflected a stranger, darker world. Newton wrote of a universe teeming with life – 'so may the heavens above be replenished with beings whose nature we do not understand.'[11] He never published these thoughts. But it was a speculative vision shared by another seventeenth-century cosmologist, the Dutchman Christian Huygens, who spoke of other planets having 'their Dress and Furniture, and perhaps their Inhabitants too as well as this Earth of ours.'[12]

These are strange thoughts. Perhaps in removing man from his privileged position, both Newton and Huygens felt they must insist that he was not alone, that nature was full of alien beings of whom we knew nothing. Certainly Newton never derived the bleak, conclusive certainty from his own work that we do. And that has been our tragedy.

'I do not know what I may appear to the world,' he said to

his nephew towards the end of his life, 'but to myself I seem to have been only like a boy, playing on the sea shore, and diverting myself, in now and then finding a smoother pebble or a prettier shell than ordinary, whilst the great ocean of truth lay all undiscovered before me.'[13]

Newton and Huygens were to be the last scientists free to speculate about the possibility of a warm, inhabited universe.

'The eighteenth century dawned bleakly,' writes Freeman Dyson, 'under a heaven grown empty and dead. Cosmology, ever since that time, has concerned itself only with an empty and dead universe. When Newton decided to suppress his youthful visions of the cosmos, he was only doing what every good scientist is supposed to do, abandoning without mercy a beautiful theory which turned out to be unsupported by experimental facts.'[14]

In addition, much of Newton's working life was taken up with the very forms of knowledge of which he is taken to be the prime destroyer – alchemy and astrology. Surveying his life as a whole what we have come to call Newtonian mechanics is revealed as just one part of his mighty imagination. Like Shakespeare before him he spoke with many voices.

Newton was a man holding at least two worlds together in his mind, two worlds which we now believe to be contradictory: one was magical, the other scientific. 'Newton was not', wrote the British economist John Maynard Keynes, 'the first of the age of reason. He was the last of the magicians, the last of the Babylonians and the Sumerians, the last great mind which looked out on the visible and intellectual world with the same eyes as those who began to build our intellectual inheritance rather less than 10,000 years ago.'[15]

And the clarity of his science extended to his awareness of his own role. Seeing precisely where he stood, he made clear the importance of the gap between science ('philosophy' in his terms) and religion: 'We are not to introduce divine revelation into Philosophy, nor philosophical opinions into religion.'[16]

It was the distinction of the age and the distinction by which we now live. It says there was no value in the world and there is no science to be applied to the realm of the transcendental. The two are separate. We cannot look at the heavens and find God, nor can we apply the laws of motion to the afterlife. Man was divided in two. Yet, even here, Newton was, as Keynes suggests, only ambiguously modern. For he saw himself not as a discoverer but as the latest prophet from a long tradition. He was simply the deliverer of God's truth to his generation. He was nothing new.

But, we now know, he had unleashed a transformation of the human imagination. His statement of how little he knew, of himself as a curious boy on a beach, can be taken as a realistic awareness of how much science was still left to be done or as a statement of real ignorance of the nature of the world. From our perspective we can choose either, for this protean magus and philosopher contains both.

But the world chose science rather than magic because it thought it was 'true', because it worked. What must never be forgotten, however, is that it *was* a choice, we adopted a particular perspective, a perspective which, to Newton, would have been only half the picture. The other half would have been the spirit world of alchemy, sorcery and demons.

In making this choice we say something about ourselves, about the type of truth we require of the world. The philosopher Ludwig Wittgenstein was to point out, after two and a half centuries of the apparently irresistible progress of this partial Newtonian model and ideal, that the fact that we *can* see the world in terms of Newtonian mechanics tells us nothing about the world; that we *do* see it thus tells us everything. And, because we do, from 1700 onwards the human imagination was convinced that nature could be fully understood as a series of differential equations, as an algorithmic compression.

So the Newton we chose – *our* Newton – made us. Newton the mechanic of the cosmos first distilled the essence of our modern conviction that science could be – must be – utterly effective. Its glory was its completeness, its totality and its efficiency. Science's power and the magic of its numbers were dazzling, overwhelming. Progressively they would exclude all possibility of competition, of alternative explanation. Many figures have been held up as 'makers of the modern world', but only Newton really earned the title.

For all their clairvoyance, however, Newton and Galileo did not address the most pressing crisis created by their work. Perhaps because what we now know as philosophy was not their craft. Perhaps because they believed their own modifications of the religious view were adequate: God, for Newton, had merely set the mechanism in motion, for Galileo he had simply written two separate books, one concerning man's salvation and one concerning nature.

But another seventeenth-century man saw the problem posed by science more clearly then either. If Galileo invented the modern, and Newton subdued the Heavens, this man wrote the rules by which we chose to live. He saw that the success of science demanded a new investigation into the nature of knowledge. He was a restless, curious Frenchman who felt the cold on his extensive travels in Northern Europe and who saw the cardinals gathering together in Rome during the years in which the Church was warily assessing the meaning of the new science. He was a man physically and intellectually at the centre of things.

On 10 November 1619, he rested at the town of Ulm, the birthplace of Einstein, *in* a stove, say some versions, or in a room heated by a stove, say others. As the warmth penetrated his chilled body he had a vision of the unification of all science. Later he went to bed and dreamed three dreams. In the first he was being spun round by a whirlwind and frightened by ghosts. He thought he was to be presented with a melon from a distant land. The wind calmed and he woke up. In the second dream he saw thunder and lightning in his room. The third dream was quiet, he read the poetry of Ausonius – 'What path shall I take in life?' ran one line. A man appeared and quoted more poetry. The dreamer searched for his book but it had vanished. The dream ended.

He made a pilgrimage to Our Lady of Loreto and prayed. The night's visions, he concluded, had been a foreshadowing of his destiny to unify all scientific knowledge and, eighteen years later, that is what René Descartes, the new dreamer of our age's dream, did.

3
The humbling of man

A dog might as well speculate on the mind of Newton.
Let each man hope and believe what he can.

Darwin[1]

Galileo died in 1642, the year of Newton's birth. He had been born in 1564, the year of Michelangelo's death as well as Shakespeare's birth. The great men of the first age of science were magically linked.

'I commend these facts', wrote the twentieth-century English philosopher Bertrand Russell with arid, liberal humour, 'to those (if any) who still believe in metempsychosis.'[2]

Russell was an apologist for science. He turned philosophy into its handmaiden. He would have thought metempsychosis – the movement of a soul from body to body, which could have made Newton the reincarnation of Galileo – as nothing more than an absurd old superstition. This chain of great souls was a coincidence, no more. Then why does it intrigue us? Because there is far more in heaven and earth than is dreamed of in Russell's philosophy.

Galileo had died blind. Having projected man's gaze across the heavens, he ended his days locked in the dark, narrow prison of himself. It was an irony he, of course, as a modern man noted with wry anguish. Newton died in 1727, in full possession of his senses. A friend was disturbed that he had not asked for the Last Rites, but consoled himself with the strange observation that, 'It may be said his whole life was a preparation for another state.'[3]

Newton had seen himself at the last as a small boy on a beach; Galileo as a shrivelled, broken man. They had achieved everything and yet nothing. They were humbled men. They had changed the world, but not their own human condition. They had created a hunger for a food of which none could yet conceive.

That food would have to provide sustenance for the human soul in the face of the cosmic loneliness which science had revealed as its 'true' condition. What was needed, now that the universe had changed for ever, was a bridge between the inner, human reality and the outer, cosmic order they had revealed. Newton himself, as the last great sorcerer, had no need of this bridge. His mind held the two worlds in balance. Galileo was an enthusiastic devotee of the *Scienza Nuova*. Getting on with the truth was, to him, an obvious task.

But what was the truth? The very idea had suddenly become slippery, evasive, uncertain now that ancient authority had been undermined. The word itself had begun to take on the first overtones of the modern ironical inflection. We no longer use it except when, embarrassingly, our vocabulary lacks an alternative. Modern embarrassment with big ideas is one of the more curious legacies of science.

The slipperiness of truth had to be contained if science was to have a foundation as secure as its Aristotelian precursor. And containment was to be Descartes's task.

He was born in 1596 and died in 1650, so he managed, just, to be a contemporary of both Newton and Galileo. He was to write the rules, the script for science, though he was never to see the full flowering of the knowledge he was to codify in the conclusive revelation of Newtonian mechanics.

If Newton's destiny was to be the last of the old sorcerers, Descartes's was to be the first of the new philosophers. He is routinely called 'the father of modern philosophy' and this might be right, for he created the terms of almost all future philosophical debate. I hope to show later how, I believe, only in our time has his reign finally come to an end.

Le Discours de la Méthode pour bien conduite la raison et chercher la vérité dans les sciences appeared in 1637. It was a preface to a set of three essays: one on optics, one on meteors and one on geometry. Descartes was an important scientist and mathematician as well as a philosopher and it was the science that formed the substance of the book.

But it is the preface and his later philosophical works for which he is best remembered and which is embodied now in all our lives. For Descartes had embarked on a project that was to define the spirit of the coming age as precisely as Newton was to define its mechanics.

'It turned out', the philosopher Alasdair Macintyre has written, 'that Descartes had woven into a single rational system some of the dominant themes of the next age, in its life as well as in its thought: the isolated individual as self-sufficient in knowledge and action; the ideal of mechanical explanation; the reduction of God to the status of a guarantee that the gaps in rational argument can be filled, and the actions of individuals harmonized; the dualisms of reason and the passions, and of mind and matter. Cartesianism is the new consciousness expressed as doctrine.'[4]

'I cannot forgive Descartes,' wrote Pascal in endorsement of his centrality but denial of his wisdom. 'In all his philosophy he would have been willing to dispense with God. But he had to make Him give a fillip to set the world in motion; beyond this, he has no further need of God. . . . Descartes useless and uncertain.'[5]

Pascal was again seeing the terrible truth of his age. He had seen that Descartes had done everything except secure the position of God, and thereby the meaning of man.

To understand just how radical Descartes was, it helps to look at a slightly earlier attempt to lay a philosophical foundation for the new science. Francis Bacon was an English lawyer who embarked on a massive work – *The Great Instauration* – intended to provide a complete method for and justification of the unlocking of nature's secrets. He died in 1626 before completing his task, but his design and intentions are clear.

Bacon was an experimentalist, so he rejected the abstractions of Thomist scholasticism. His attitude anticipated the practicality of Newton in grinding his own lenses. His distinctive innovation lay in his acceptance of the modern view that we cannot simply *reason* our way to the truth. Experiment and observation are also required; reason alone could not assure us of the validity of our conclusions. In this he embodied a certain robust practicality which was later to become regarded as a national characteristic of the English when it was realized in the Industrial Revolution.

Yet Bacon did not reject the essential outlines of the Aristotelian world view. Indeed, in his fascination with the multiplicity of discrete facts to be catalogued in the world he can be seen as an utterly faithful disciple of Aristotle. In this he represents an almost precise midpoint between the old wisdom and the new science. What he proposed was a more realistic project of knowledge than that of the scholastics, but one that was considerably less radical than that of Descartes.

Bacon conceived of science as a massive, communal effort of induction. Facts about nature would be accumulated and, from this great body of knowledge, general laws would be formulated. The project was practical in that it was entirely aimed at improving mankind's lot. Bacon may, therefore, be said to have codified the idea of technology and, in the communal nature of his conception, defined the constitutions of the great scientific societies that were to be founded in the eighteenth century.

Low temperature afflicted him as it would his successor, Descartes. Curious and experimentally inclined to the last, he died of cold after stuffing a dead chicken with snow to see if low temperature would preserve the meat. It would have done.

Bacon had responded to the increasingly obvious success of pre-Galilean, Renaissance curiosity about the world by attempting to formulate what would have been a middle way between the defensive abstractions clung to by the Jesuits and the aggressive rationalism inspired by the power of the new knowledge. The style of his thought is that of many modern scientists – realistic, practical, suspicious of abstraction and optimistic. He remains a hero of the scientist who aspires primarily to virtues of forthrightness, realism and practicality. The Baconian gets on with the job.

But he was premature in that he was explaining and justifying the colourful curiosity of the sixteenth century rather than the blinding revelation of the seventeenth. The result of this is that, however homely and sane his method seems, there is an important sense in which he was not addressing the real challenge of science to the prevailing Christian order. For the Baconian, inductive style cannot *ultimately* be Christian – at least not to the refined inheritors of the scholastic tradition. Any monotheistic religion must, by its nature, be driven towards the unification of knowledge. One god means a single underlying pattern. Certainly

Bacon was in pursuit of laws, but his first interest was in the vast project of cataloguing the facts.

What did not concern him was the more complex issue of the deeper implications of science for the foundations of our knowledge. This must be the true concern of a monotheistic culture that had inherited the tradition of Aquinas and it was the genius of Descartes to see that this was the one issue that was to define the world of the future. For the real mood of the modern world is not the bluff certainty of Bacon, it is the refined disquiet of Descartes.

The *Discours* describes four essentials of the scientific method: accepting only what is clear in the mind, breaking down large problems into smaller ones, arguing from the simple to the complex and, finally, checking. Simple enough, but what lay behind this simplicity was the most radical response to his age's confusion, scepticism and doubt. For, also in the *Discours*, was the Cartesian motto: 'Cogito, ergo sum.' I think, therefore I am.

The problem that lay behind the energy and excitement of the *Scienza Nuova* was the problem of knowledge. Bacon had solved it to his satisfaction with a reversion to a robust Aristotelian empiricism. But the very fact that a new solution was required was the real point. For example, no solution was necessary to a theoretically perfect Catholic. Looking out on a Thomist world with the eyes of faith, he could reasonably claim to 'know' everything. In medieval Aristotelianism knowledge was arrived at by proceeding from first principles. Effects were shown from causes and its propositions must, necessarily, be true if they conformed to logic. The logic ensured a theoretical immunity to challenge from experiment, observation or from dissident, individual reasoning. It was the truth of the scholar, self-justified by its own coherent existence. There were, of course, many disputes, but the underlying emphasis was clear – there was a divinely inspired truth towards which our reason could guide us.

The centuries of Renaissance, Reformation and discovery as well as the ensuing turmoil of European wars steadily overturned all such certainties. The old order was being subject to the violent changes that now provide the familiar basis of our schoolroom histories. Set against these changes, science may be seen as a rather fragile, cerebral development. But, to the percipient, it was the distilled embodiment of them all. Wars may come and go, but this time the wars signalled a fundamental change in the human soul far more deadly to the old order than any number of armies and more lasting than the passing demands of any political, commercial or territorial interests.

'And new philosophy', wrote the English poet John Donne, 'calls all in doubt.'[6]

In the simplest terms, as I have shown, the universe was found to be other than belief and authority had said. But, even more importantly, the basis of knowledge had gone. By definition the old system was so complete, so dependent on its own internal coherence, that the destruction of its form meant that its method was destroyed as well. It was like waking up to find that your home town had been a film set all along, a two-dimensional illusion of wood and canvas. And Descartes was about to apply a match.

To understand this is to understand much, perhaps all, of what has happened since. Certainly it is to understand the dark, pessimistic side of the modern world which has defined so much of the twentieth-century imagination. For ours has been the exhausted, disillusioned phase of the Cartesian universe after the initial optimism inspired by scientific and technological triumphs had ebbed. In the crisis of the loss of a coherent framework for what we know or can know the modern world was created.

But what, we might now ask, was the problem? Science seemed to show that we could work out the truth about nature. All right, it was a different truth from any that had gone before, but so be it.

The problem was, simply, that this new truth appeared to be based on very little. When Galileo placed his eye to the telescope and saw the impossible, he had the courage and genius to believe that he alone had seen something fundamentally new. The idea of novelty, of perpetual change, discovery and innovation, had entered the world as had the modern idea of the scientist as capable of solitary understanding. But where was this knowledge coming from? It came only from Galileo, one man among many.

'O presumptuous man!'[7] cried Pascal.

In contrast, consider the knowledge and wisdom of the Church. This was the work of centuries and of countless saints, visionaries, scholars, geniuses, popes and priests. There is no basis for suggesting that Galileo was a greater genius than Aquinas or St Paul. Why, then, was he more definitively right? Why was his spark of reason worth more than the authority of thousands?

'On what shall man found the order of the world which he would govern?' asked Pascal. 'Shall it be on the caprice of each individual? What confusion! Shall it be on justice? Man is ignorant of it.'[8]

Pascal was noting the great modern irony. Our knowledge allows us to do so much, and yet it also exposes us as small, accidental and ignorant. Man alone is but a feeble foundation for truth.

In the characteristic shallowness of our day, we might say that Galileo was more right than Aquinas because he came later and so he knew more. Today if you challenge scientists when they claim to be close to a Theory of Everything, this is the defence they employ. Our new experimental and observational capacity *must* take us closer to the truth than the scientists of the past.

But this is to read our idea of the progress of discovery backwards into history. This idea was invented by Galileo and his contemporaries. It would have no meaning at the time. Progress was not a conceivable basis for saying that Galileo was more right than Aquinas.

Or, perhaps, we might say that Galileo had the technology in the form of the telescope. But others had looked through this instrument. Only Galileo had seen the entire universe fall into place and few could understand why it had done so, the rest had to take it on faith, a new faith, much as today most of us are obliged to take relativity or quantum mechanics on faith.

Perhaps it was clarity of observation. But there had been clear observers before and they had been proved wrong. Besides, a scepticism born of the new age warned against taking anything as evanescent as sense experiences too seriously.

So, on the face of it, the authority of *Scienza Nuova* was based on appallingly shaky foundations. It worked, sure enough, but so had other systems at other times. What was it about this horrifically effective infant that made it distinctively different?

Part of the answer can be found in the context from which it sprang. The age was jaundiced. The turbulence of the times had made men suspicious of authority and reason. Scepticism was a potent force politically as well as philosophically. Dogma meant more than just belief, it meant power. Transubstantiation had been the dogma the Jesuits had chosen as their test of faith and the assertion of their power. But the human reality of its adoption was not quite the sublime gift from God that would have served their purposes.

The Council of Trent in 1545, which had laid down the orthodoxy of the Counter-Reformation, had not been the perfect restatement of the faith the Church had intended. And, in modern terms, its imperfections were 'exposed'. A written history of the council by Paolo Sarpi was handed over to the Protestant English in 1618. It was the most diplomatically devastating book of the age, revealing the council's deliberations as a perfectly ordinary chapter of political manoeuvrings. The primacy of transubstantiation was *negotiated* into existence, according to Sarpi, as if the divine entered into matter on the basis of a trade deal. Protestant Europe, more intent on the private salvation of individual souls than on the salvation of the Roman Church, was delighted. Scepticism as an attitude to religious authority and the world was endorsed.

The fact that a book could do such damage was also singularly appropriate. Protestantism had insisted on a return to the primacy of the Bible and this had inspired the centrality of the image of a man alone with a book, searching for his own truth without the aid of a pompous authority that would have preferred to keep him in ignorance. The idea of the book became a symbol of the revolt against authoritarian Catholicism. Indeed, it has been said that, without printing, the Reformation would have been impossible. It was a revolution of communication and persuasion, of words written and printed on paper. In our day the mass media have inherited something of the same aura of virtue – their rationale is the right of the individual to find out for himself and to make his own decisions. Such knowledge subverts authority. It demands access and evidence.

The violence of the age in which modern science was born acted as a midwife to the new way of knowing. As science undermined old authorities, it also encouraged new societies. The age of religious wars came to an end with the Treaty of Westphalia in 1648 and the age of wars of economic nationalism began with the Anglo-Dutch conflict, a war that coincided almost exactly with the years of Newton's greatest creativity. Economics imposed new criteria on the conduct of policy. Practicality and Machiavellian pragmatism took the place of the mystical justifications of the past. Man became a more calculating, prudent being.

The historian R. H. Tawney described the transformation: 'From a spiritual being, who, in order to survive, must devote a reasonable attention to economic interests, man sometimes seems to have become an economic animal, who will be prudent, nevertheless, if he takes due precautions to assure his spiritual well-being.'[9]

So prudence thrives but Truth, in such a culture, becomes a lonely task. Individualism can be heroic or it can be merely baffled. There is no guidance that can be offered. Where, then, could the honest man find Truth? Truth could not be in the gift of the bureaucrats of the Church.

The honest man may conclude that what he sought was still a Christian truth, but what was that? After Luther there had been a series of variations on the Protestant theme, all apparently based on impeccable scriptural authority. All were 'Christian truths' – but how could they be?

Yet, amidst these conflicting voices, there was also this system, this *Scienza Nuova*, which was, apparently revealing some new truth. It worked, but on what basis? Abandoning the authority of centuries for that of a wayward genius such as Galileo must have seemed like a poor exchange.

But, Descartes saw, it was an exchange that would have to be made. The old physics and cosmology were discredited. Science demanded a new basis for knowledge. It did not demand it of scientists, of course. Then, as now, they were content to carry on with their work because they accepted the adequacy of its internal logic. But, however hard-headed the expression of such a view might be, it remains a metaphysical position. Scientists who insist that they are telling us how the world incontrovertibly is are asking for our faith in their subjective certainty of their own objectivity. It represents a fundamental philosophical challenge which Descartes and, in one way or another, every philosopher since, has tried to answer.

Descartes's method, in spite of the visions in the stove that attended his inspiration, was the application of hard philosophical scepticism. This he took to what he thought was the ultimate extremity. He was a believer, but he saw that belief alone could not produce the clarity and lack of ambiguity he, as a scientist of the new age, required of the world. He could not rely on his senses, he decided, because they could be deceived. One even had to take into account the possibility that they were being systematically and completely deceived by some demon. What was left? His own sense of himself, his awareness of his own existence. Cogito, ergo sum.

It is here that the *Discours* is at its most recognizably modern. For its central theme is not the traditional philosophical one of the objective pursuit of knowledge of the world, it is the mind of Descartes. Here, in the interior of this skull, is where the drama takes place. We must ask questions of ourselves before we can expect answers of the world. Autobiography is the beginning of philosophy. He was the father of modern philosophy, but he was also the father of modern self.

As Bernard Williams says, the book 'displays its author not so much as an object of human interest to himself or others, but rather as an example – though a genuinely existing, particular, example – of the mind being rationally directed to the systematic discovery of truth.'[10]

Descartes's mind, as a mechanism, had become part of Western history. Even here he anticipated our age: Albert Einstein's brain was later studiously preserved as if in its material reality a clue to the universe it had encompassed in life could be found.

But Descartes, I repeat, was a believer. He had found a basis for certainty within himself, but he had to join this to his faith. The only thing outside his mind of which he could be as sure as he was of his own existence was God. Descartes's God was the foundation of the delicate structure of the self. He supported the one certainty of his own thought.

At this point the legacy of Descartes shows a crucial similarity to the legacy of Newton. Just as Newton was a sorcerer as well as a scientist, but we chose his science, so Descartes was a man of God as well as a modern philosopher. We chose the philosopher. In both cases the choice seems to have been made because the balancing reality – sorcery or faith – seems to have been too feeble to co-exist with science.

In Descartes's case it is not difficult to see, as Pascal did, that his is a very different God from that of the Jesuits; more ominously, it is not clear that such a being is necessary to support the microscopic fragment of unsceptical belief that Descartes had permitted himself. Later philosophers pointed out that his 'cogito' – I am thinking – should, according to his own method, have been more accurately expressed as 'cogitatur' – there is some thinking going on. How could he, after all, posit the existence of an 'I' from the mere phenomenon of thought? The 'I' of the cogito is a kind of grammatical error and its God a mere syntactical necessity. Descartes's scepticism was as nothing compared with what was to follow once his God had been discarded.

This crucial fragility of Descartes's position was defined later by others. Yet he knew himself that he had placed man in an odd, almost dreamlike state. He saw himself as 'half way between being and nothingness', precariously held in existence by the will of God. Again this defines him as the philosophical father of the modern world. Remove God and you have modern, existential uncertainty. Indeed, the primary philosophical work of Jean-Paul Sartre, the prophet of existentialism, is called *Being and Nothingness*. Once Descartes had defined the fragility of modern knowledge, the Western imagination could not recover. Cartesian man is possessed by an enfeebling sickness that leaves him dreaming on the edge of oblivion. Descartes infected him by simply pointing out how limited were the available certainties beneath the corrosive gaze of honest scepticism. No understanding of the present can hope to work that does not include him at its heart.

Yet Descartes himself had entertained the pious hope that his own doctrines would become the orthodoxy of the Catholic Church. After Galileo's trial, however, he lived in fear of persecution for what he wrote. He saw how far he had gone, how far he thought any honest man would have to go in search of the truth. What he did not see was that he had placed epistemology – the study of knowledge – at the centre of the philosophical stage and it would not be displaced for two centuries.

In suggesting an inner self-awareness as the basis of all knowledge, Descartes had established the dualism which was to remain the underlying belief of all scientific civilization. In effect, he divided us from our bodies, reason from the passions, mind from matter. Our true identities resided in our mind or soul – the seat of which, Descartes concluded, was the pineal gland in the brain. The body that appeared an indispensable companion to this self was, in fact, part of the phenomenal world that was not us. Animals, he concluded, being bereft of any such self-awareness, were merely bodies and thus could be understood as machines.

It is impossible to overstate the importance and centrality of this belief. At an amazingly early stage in the development of *Scienza Nuova* and the Church's crisis of authority, Descartes had seen to the heart of its deepest implications. Galileo had celebrated the power of reason to overcome the often deceptive and inadequate evidence of our senses. But Descartes saw what this meant philosophically as surely as Newton was to grasp its implications scientifically.

Science trapped us all in our private reasons. It divided us from our world, locked us in the armoured turrets of our consciousness. Outside was an alien landscape which was either illusory or meaningless, inside was the only possession of which we could be sure – the continual, anxious chattering of our self-awareness. Our souls were removed from our bodies.

It was this extraordinary prescience of the man that makes so much in Descartes seem so familiar. The Cartesian insight has clung tenaciously to our imaginations. One of the earliest poems – *Whoroscope* (1931) – by the twentieth-century Irish writer Samuel Beckett was a savage, uncomfortable lampoon of the life of Descartes and, in his ensuing novels and plays, the chilling, imprisoning implications of the great 'Seigneur du Perron' haunt every word. For modern man, alone, self-created, self-defining and baffled by the world, was born in the *Discours*. It is, perhaps, Descartes's greatest misfortune that the 'religious bridge' – the God that held him in being – which he had constructed to bring him back to the world from the furthest reaches of scepticism has been stripped from his legacy, leaving scientific man alone on the farther shore.

But the modern world was to have a long, hard birth and the

first effect of the new science was not the sense of isolation and loneliness that was to wash ashore in the bleak, cold visions of Beckett. Indeed, the first effect was stunned gratitude.

'Nature and Nature's laws lay hid in night:
God said, *Let Newton be!* and all was light.'[11]

The words are those of the eighteenth-century English poet Alexander Pope. The lines are often quoted as simple celebration, but there is more to them than that. There is an explanation of the new awareness. Nature, in the couplet, is a system with laws. The word is crucial and perhaps ambiguous. Newton had realized that such laws were impossible in Aristotelian physics – matter simply had intrinsic qualities, no real generalization was possible. But the image we derive from the universe of laws is that of a mechanism obeying rules from outside itself. The faithful may say God devised those rules and left the mechanism to run. Is he also subject to those rules or can he change them at will? Descartes thought he could; his successor, Leibniz, did not, believing instead in a 'pre-established' harmony which provided a rational basis for all creation and with which God could not interfere. Either way, God was a good deal more remote than he had ever been before.

And why were these laws hidden? Pope suggests that Newton was sent like some second Christ to reveal them to us, to bring light to our darkness. God had decided it was time for us to know. It was an interpretation that would have been sympathetic to Newton, the last sorcerer himself.

But, as in Descartes, this suggests a fatal weakness in the position of God. First, if nature was simply running on timeless and universal laws, as Newtonianism suggested, then God as a constant presence was not obviously necessary. Secondly, if such was the case then God's intervention or even his production of Newton was hardly necessary either. In a weak sense Newton may have celebrated the magnificence of God by demonstrating the overwhelming order of his creation. But, in a much stronger sense, he had demonstrated the power of specifically human reason, unaided by God. Man could now see immense distances, he could forecast the future, he could understand what he could not experience. Was not Newton's real achievement to turn men into gods?

Again Newton appears as an immense focusing lens of history. Not only was his thought the culmination of all *Scienza Nuova*, but he also seemed to represent a moral, philosophical and cultural climax.

In one sense men had already begun to turn into gods in the Renaissance with the emergence of heroic individualism. The greatest artists of the day were celebrated as, at least, demi-gods. In the mid-sixteenth century Vasari had written his book, *The Lives of the most excellent Italian Architects, Painters and Sculptors*, in celebration of Michelangelo, Leonardo and Raphael as men of historically unprecedented greatness. At last, Vasari claimed, the ancients were being surpassed, at last modern civilization was scaling the highest of all peaks. About the same time the Florentine goldsmith and sculptor Benvenuto Cellini had written his autobiography, an astonishingly arrogant, vain and passionate assertion of his own genius against the world. No book more vividly summarizes the sheer individualistic confidence that characterized the masters of the new age.

With the advent of science this proud individualism was supported by the idea of the lonely scientist who was able to construct the cosmos on his own terms. And, with Newton gazing into the sun, came the full realization that the mind of one man could contain the entire universe. The experimental basis of the new science – the belief in the stable, comprehensible, objective reality of a world which we could perceive and analyse – appeared to be proved by the way Newton's mathematics could be realized in the heavens. And, of course, Newton endorsed the failure of the belief that truth could be possessed by authority alone.

So, by 1700, the primary elements of the scientific universe were in place. Just over 150 years before the earth had been the privileged centre of creation, now it was a fragment in an infinite cosmos, scattered with random flecks of matter that danced in mute, unconscious obedience to changeless laws enacted in absolute time and absolute space. God had retreated to a position outside this system. He was clearly not 'up' in Heaven any more than Hell was 'down' below. Aristotelian space could encompass such concepts, but Newtonian space was uniform and without direction. It was, like all else in his system, devastatingly neutral. 'Up' and 'down' were our own invention, underwritten by nothing in real space.

Meanwhile, mathematics had become the cold language that alone could delineate these hard truths. If the numbers of the Newtonian calculus showed that the words 'up' and 'down' had no absolute meaning, then where was Heaven and where was Hell? They were relegated to the geography of metaphor. The acceptance of a degree of metaphorical understanding of the scriptures in Aquinas had widened to include all the elements of simple belief, potentially even belief itself. It was to widen still further.

The new Book of Nature was written in numbers. Mathematics, indeed, seemed to have taken the place of scripture as a means of describing the world. Newton had invented his own – the calculus – and used this wonderful abstraction to map the cosmos.

But what was mathematics? Was it in the world or was it simply a means of measuring and counting the world? The success and universality of the Newtonian method suggested the former. The universe appeared to be made out of numbers. Newton had discovered them, not applied them, like a ruler, to reality. The debate between mathematical realists, who believe numbers are in the world whether we know about them or not, and various types of instrumentalists, who insist we simply apply them to the world, continues unabated today. In 1987 when the University of Tokyo calculated the value of pi to 201,326,000 digits, were they discovering those decimal places or inventing them? We know how to persuade these strange things to perform fabulous feats, but we do not know what they are.

The imaginative effect of the realization of the power of number was the resurrection of an alternative form of classicism to that of Aristotle. Plato was reborn with the calculus.

Platonism was the original 'idealistic' philosophy. It states that our senses provide us with only a limited understanding. The forms we see about us are but distorted impressions of a world of ideal forms. Mathematics was an inkling of such an ideal order. 'Let no one destitute of geometry enter my door,' Plato had said. The forms of Euclid's geometry did not exist in the world – there was no such thing as a point with no dimensions, for example – but they lay beyond it, in an ideal realm from which our senses, though not our reason, were excluded. In Pythagoras the view had attained its purest, transcendent interpretation. Number underlay all things and Pythagorean theorems as well as his discovery of musical intervals based on simple ratios appeared to prove it. But this ancient faith never recovered from a more painful discovery – irrational numbers like the square root of two. The inadequacy of the Greek mathematical model undermined their belief in the transcendent reductionism of number.

But *Scienza Nuova* revived a number of elements of Platonism. Both Newton and Galileo worked on the basis of 'ideal' conditions – frictionless motion or bodies uninfluenced by external forces – and both relied on mathematical systems. Their laws were ideal simplifications, references to an underlying, invisible reality.

There was a more poetic parallel that appeared most vividly in the work of Johannes Kepler and in the temporarily blinding experiment of Newton. This was the worship of the sun. Ancient Platonism's insistence on the imperfection of the world of our senses had attributed perfection to the sun. So the heliocentrism of Copernicus and Kepler also tended to elevate Plato above Aristotle. Finally, the Platonists belief in perfection was more comfortable in an infinite universe and this was denied in Aristotle's cosmology. But infinity was implied in Newton's laws – he was a man, it had been said, who had no fear of the infinite. Plato had thus superseded Aristotle as the ancient model of wisdom.

It is important to be aware of this Platonism attached to early science because it neatly reverses certain preconceptions. Platonism is mysticism, Aristotelianism is common sense. Science is Platonic in inspiration yet it is more commonly associated in our minds with the idea of common sense – indeed, one of the more celebrated definitions of science is that it is 'organized common sense'. But I believe the Platonic roots of science point to a deeper truth – that science itself is a form of mysticism. This would be unpalatable to all the popularizing apologists of science precisely because they must cling to common sense as a basis for their propaganda. But, as I have said, common sense can see the heavens through the eyes of Ptolemy; it takes mysticism to see them through the eyes of Newton.

Beneath all the innovation, confidence and celebration attached to the new science, there remained the essential coldness of the scientific and mathematical vision. If the world was really made of numbers, then we inhabited no more than a giant calculating machine. We may take consolation, with Descartes, in our own consciousness, but it was no more than a tiny, mortal flicker in the machinery of nature. Furthermore, these laws and these numbers appeared logically to be operational with or without us. Efficient cause and effect ground on through the ages and all the petty concerns and struggles of our lives could alter nothing. We had been displaced from the centre of universe, from our privileged place, from our meaning.

Even more alienating was the way the cosmos of number robbed us of our freedom. The clear implication of science was determinism – the belief that everything was eternally fixed. The future was contained in the present which was predestined by the past.

The perfect expression of this view came from the French mathematician Pierre Simon de Laplace in his *Philosophical Essays on Probabilities*.

'An intellect,' he wrote, 'which at any given moment knew all the forces that animate Nature and the mutual positions of the beings that comprise it, if this intellect were vast enough to submit its data to analysis, could condense into a single formula the movement of the greatest bodies of the universe and that of the lightest atom: for such an intellect nothing could be uncertain; and the future just like that past would be present before its eyes.'[12]

It was the urbane Laplace who, on presenting a copy of his book to Napoleon, was asked why it contained no mention of the Creator. 'I have,' he replied, 'no need of that hypothesis.'

Laplacean determinism is an inescapable conclusion of the mechanics of the scientific view and it remains the underlying belief of most scientists. If we can know every cause, then, logically, we can know every effect. The future is knowable if we can process all the information about the past. It is, therefore, fixed and unavoidable. We cannot escape this by fleeing into our

Cartesian souls for we too must be made of numbers and these numbers must be subject to the same iron logic of cause and effect. We do things because of other things and we are joined to the whole universal chain of causality. Free will is an illusion born of our ignorance. Science tells us that we can know everything and, therefore, that we can be utterly imprisoned.

The eighteenth century was the age of the triumph of reason and the immense confidence that stemmed from the conviction, inspired by Newton, that the human intellect could achieve anything. But, behind this confidence, lay religious and philosophical terrors. Reason may have triumphed, but the fruits of victory were proving bitter. Science had abolished free will and established a lonely, brutal universe.

After Descartes, as I have said, it became clear that the philosophical response to such crises lay in the solution of the problem of knowledge. Only then could we move on to the more sensitive subject of value, morality and purpose.

The overwhelming philosophical impact of science was the separation of knowledge from value. Indeed, this seems to be what ensures its success. For science is, of necessity, dynamic. It requires always the possibility of experimental refutation and a permanent process of scepticism about its own findings. But, if we attach a value to one particular view, then either the process is paralysed or the value is vulnerable to overthrow. Say, for example, we state that Newton had interpreted the mind of God, as Pope implied. What can we say when Einstein appears to expose the incompleteness of Newton? The nature of God's mind cannot change with every new theory. Science is always restless and always destructive of any attempt to freeze its conclusions into a more than scientific truth.

Being human, scientists attach values to their theories all the time, but the point is that the underlying experimental assumption must be that no one theory is privileged, however attractive it may be. The scientific world is not made to satisfy us and nor can it answer anything but scientific questions.

Even Descartes himself, the pious man, saw that drawing conclusions about God from his creation was invalid. He believed that the anti-mechanists who insisted on the old Aristotelian view of final causes in nature were being impious by attempting to second guess God. Final causes were simply not within the realm of human reason. We had this mechanism and that was all; God was remote.

And from remoteness to absence is but a short step. The cold mechanism of the universe had replaced God's benevolent creation. Science made it progressively more difficult to study nature in order to understand God and derive our values from his work.

There is a complex pressure at work here which it is important to understand. On the one hand science had demonstrated a new order in nature that would seem to endorse the existence of a guiding, creative hand. On the face of it the old 'argument from design' should be strengthened by the new forms of knowledge. This argument said simply that the complexity and adaptive perfection of nature made it unthinkable that it had not been consciously designed. Science revealed greater complexity and even more perfect adaptation – surely, therefore, it provided even more beautiful examples of the divine designer. What could be more compelling evidence of design than the superbly robust yet delicate balance revealed by Newton?

Such an attitude has always underpinned the work of the believing scientist. It found its most celebrated and elegant expression in the writing of the eighteenth-century theologian, William Paley.

'In crossing a heath,' he imagined, 'suppose I pitched my foot against a *stone*, and were asked how the stone came to be there; I might possibly answer, that, for anything I knew to the contrary, it had lain there for ever: nor would it perhaps be very easy to show the absurdity of this answer. But suppose I had found a *watch* upon the ground, and it should be inquired how the watch happened to be in that place; I should hardly think of the answer which I had before given, that for anything I knew, the watch might have always been there.'[13]

The watch might be a tree, a bird or a man – all equally improbable unless designed.

On the other hand science did appear to have explained and simplified some very complex systems. The bewildering cycles of Ptolemaic observation had been transformed and understood. The hard, atheist scientist might simply take the view that, in due course, trees, birds and men would also succumb to its explanatory power.

For a time it must have seemed that science could go either way. Perhaps, although it had displaced us from our privileged position, it had revealed a new type of value in our deeper awareness of the order of nature. Or it had showed conclusively that we were meaningless accidents in a cold universe.

In David Hume, the eighteenth-century Scottish philosopher, the modern form of the crisis of knowledge and value first began to achieve its full expression. Hume was an empiricist in that he thought all reason stems from experience. But he did not think experience itself was adequate for an understanding of the world. From experience we seemed only to construct a world suitable to ourselves. If we saw one billiard ball strike another and send it rolling away, what we saw were two discrete events. But then we said the first event 'caused' the other. This causality was not in the world, it was placed there by ourselves. Given such an inadequacy, it would, of course, be absurd to believe our knowledge of the world as such could lead us to God. Again the strange familiarity and mystery of causality was the paradox around which our knowledge and powers seemed to pivot.

Immanuel Kant said Hume awoke him from his dogmatic slumbers. Kant was educated and taught at Königsberg in Prussia. His slumbers were prolonged. Becoming a lecturer in 1755, he worked on minor philosophical and scientific treatises until, in 1781, he produced the *Critique of Pure Reason*, the announcement of his awakening.

If Aristotle embodied the wisdom of the ancients and Aquinas the wisdom of the Middle Ages, then Kant embodied that of the Scientific Enlightenment. His system represented the most comprehensive and direct assault on the epistemological crisis of the modern world. He was to Descartes what Newton was to Galileo: a giver of laws as opposed to an explorer.

Allan Bloom summarizes Kant's achievement: 'He developed a new epistemology that makes freedom possible when the science of nature is deterministic, a new morality that makes the dignity of man possible when human nature is understood to be composed of selfish natural appetites, and a new esthetics that saves the beautiful and the sublime from mere subjectivity.'[14]

Kant was an heir of Plato, an idealist who considered that our understanding of the world was radically and unalterably deficient. All that we had was our mechanism for perception. To explain this he adopted Newton's absolute time and absolute space as *a priori* – that is to say given prior to sense experience. We examined the world through lenses of absolute time and space. They were the conditions of our seeing. We could, through our perceptions, know nothing of ultimate truth. Our knowledge was also conditioned by certain categories of experience such as causality.

On the basis of this metaphysical foundation Kant constructed a morality and a belief. We created our world as ordered by the forms of our knowledge. Within this structure we felt a moral pressure – a categorical imperative to behave always with the question in one's mind: what if all men were to behave like this? Or, more pointedly: act as if your action was about to become a law of nature. God, for Kant, was not perceivable in the world but in this ordering and moral pressure. We were being pushed towards the highest good of all mankind. But the force was defined from within ourselves.

This is the pivotal defence of the moral order against the claims and efficacy of science. Kant's system is a direct reaction to the cold, inhuman cosmology of science. Once man had been removed from the centre of the universe, the stars became meaningless. Neither they nor anything else in the cosmos gave man purpose or identity. Nor did they give certainty or knowledge. Confronted with the cold vision of doubt and indirection, Kant turned the Western mind inwards to the contemplation of its own moral and metaphysical structure.

It was the decisive intellectual effort of the modern world. All philosophy before and since can be seen to be implicit in Kant's grand system of internalized value. From one version of Kant sprang romanticism with its self-justifying, self-defining heroes and, in our day, another Kant lies behind all the varieties of existentialism and analysis in twentieth-century philosophy. It has been Kant's role, in all the many forms in which he has been recreated by his interpreters, to defend man against the devastating inroads of his own creation, science, by asserting in a new form the old religious conviction that he was, indeed, unique; that man must never be a means to an end, he was an end in himself. Value was restored to the scientific universe by its resurrection in the irreducibility of man.

But Kant's subtle, magnificent and ascetic progress towards his metaphysic was a fragile spectacle when set against the crude, explosive vigour that science had unleashed upon the world. Irreducible man was to mean little before the black-hearted power of industrialization.

In the eighteenth century modern technology, the practical incarnation of science, was born. First in England and then across Europe the Industrial Revolution took hold. The machine age began as man discovered his strength need no longer be that of his arm alone nor his speed that of his feet. Factories and railways filled out the net of knowledge which science had cast over the landscape. Richard Trevithick, Josiah Wedgwood and Benjamin Franklin were the solid, curious, energetic inheritors of the world that Descartes, Galileo and Newton had made. Blindness, alchemy, fragility and God were forgotten as the practical men took over. *Scienza Nuova* was translated into the new science and it worked. Now men were placing their hands on the levers of the machine they knew the world to be and the lives of millions were to be transformed.

Rural areas became depopulated as the peasantry abandoned the land to become the urban proletariat. The idea of the machine, a mute, efficient man-made realization of the eternal laws of science, became the dominant idea in the European mind. Machines promoted efficiency which created wealth in one vast, impersonal mechanism. They also demanded centralization and thus created cities in which the workforce that served the machines clustered around the factories that housed them.

Technology was, above all, the ultimate, unarguable assertion of science's one big claim: it works. From Newton onwards science poured out its laws and technology turned them into steam-engines, factories, cars and rockets. And the engines moved and the rockets went up. It works. Doubt amidst such profligacy of achievement was absurd and, today, this profligacy is all about us.

We may wish to say prayers to make our crops grow, but we know that fertilizers work better; we may wish to protect isolated cultures from the inroads of technology, but we know that antibiotics will stop their children dying; we may wish to walk to America, but we know we must fly.

In this context we can see the quality of the watch on the heath as an image for William Paley. The watch was a technological product. As technology expanded man began to understand and appreciate still more the power and perfection of design needed to make such a thing work. Just to make a steam-engine move required precision engineering and sophisticated production systems; consider, then, how much more was needed to make a bird fly or a man think. Man's technology was triumphant, but, surely, God's nature awarded the prize.

But the mistrust of the philosophers of the enlightenment like Hume and Kant for this idea proved prescient. Science was too mobile, restless and extravagantly creative to allow us to draw the simple conclusion that complexity equalled God. The argument from design was not to survive the dark turmoil of the nineteenth century intact. In our century, except in little pockets of resistance and in church sermons, it has not survived at all.

Time was its undoing, a neat, cruel irony since Paley had used the image of the watch as its primary defence. This was not Newton's absolute time nor the subjective time of our lives, but the time that had passed on earth. This was clearly a subject of some sensitivity to a religion like Christianity that aspired to encompass and explain all history. To support this ambition, the Bible had been used as a way of deriving a precise chronology of creation as well as a forecast of the apocalypse. Tracing back to Adam and Eve through the genealogies of scripture had, in fact, given a widely accepted date for the creation of the world as 4004 BC. History moved forward from this point to reach a climax in Christ and would at some appointed time in the future come to an end with his Second Coming. Symmetry derived from Scripture suggested to some that this would occur in 4004 AD.

The blow dealt to the old religious universe by Newton and Galileo did not necessarily cause problems for such an interpretation. The time of the cosmos might be absolute, but the earth may still have such a relatively short history. And, for the moment, few people realized that there was a *need* – scientific or religious – for a longer history.

Very suddenly, however, the need became unavoidable. James Hutton, born in Edinburgh in 1726, studied law, medicine and, finally, geology, though no science by that name then existed. From his geological observations Hutton constructed what he called, appropriately enough, a 'world machine'. This involved a four-stage process: erosion of the land, deposition on the depths of the sea, heating and compaction and, finally, the fracturing and rising of the seabed to form new land.

The first novel feature of this scheme is the idea of continual movement. Against what might be thought of as the common-sense view of the earth as a static lump of rock, Hutton insisted on this process of continual, if unobservably slow, change. It was a classic case – and one that would have delighted Galileo – of reason and observation combining to defeat the evidence of the senses. That which we had held to be the very image of solidity and stasis was, in fact, in constant movement.

Geology as a science was created by this central idea of colossal and continual change. Indeed, one modern geologist points out that the entire basis of science can be communicated in an instant by the elementary fact that the summit of Mount Everest is made of marine limestone.

But Hutton's machine held even deeper terrors than merely the realization that the very earth was moving under our feet. That we could not detect the movement could mean only one thing: it was very slow. And, if it was moving so slowly, then surely the time involved for even one part of his cycle was immense. His world machine was, he concluded, eternal. 'The result, therefore, of our present enquiry', he wrote, 'is, that we find no vestige of a beginning, no prospect of an end.'[15]

John Playfair viewed one of the 'angular unconformities' in the rocks that had inspired the theory while on a field trip with Hutton. 'The mind,' he recalled, 'seemed to grow giddy by looking so far into the abyss of time.'[16]

'Deep time' had been uncovered. The scientific imagination's relentless probing into causes now demanded the existence of the earth for incalculable aeons. Playfair's reaction was precise – he grasps the concept of deep time by employing a spatial metaphor, the abyss. Seeing into deep time he grows giddy as if at the edge of a deep space. Infinite space had now been jolted by infinite time and, in both, we grew giddy at how lost we were, how small, how insignificant. Science had shown we were lost in space; now we were lost in time as well.

But still man himself could just about remain untouched. Still he could claim his unique, divine spark, his self-aware soul that seemed so detached from the creation that science was so ruthlessly stripping to its essentials.

Even this last vestige of pride, of singularity in creation, however, did not last long. In 1859 man was attached painfully and decisively to the natural world by a scientific theory of such awesome smplicity and power that even 150 years of familiarity have not subdued its capacity to take the breath away.

Again time was the key. Why, Charles Darwin wondered, are there so many varieties of life so well adapted to their surroundings? His answer was, essentially, non-random death, better known as natural selection. Life began, an event that remained beyond even Darwin's competence to explain, and was at once subject to natural selection. The mechanism was absurdly obvious: some individuals survived to breed better than others. They thus passed on their genes to a greater number of offspring. These genes would transmit the qualities that made the parents successful and greater numbers of successful, in the sense of better adapted, organisms would be created. Chance mutations offered up the possibility of variation. Usually these would result in a less successful individual that would die off unnoticed. But, when they worked, they would allow organisms to survive and breed more efficiently. Living things usually died for a reason – because they were ill-adapted. Those that were better adapted lived and bred. Time did the rest.

The painful slowness, imprecision and chanciness of such a process immediately affront our intuition. But the only step that is really subject to chance is the possibility of mutation. All else is clear, statistical necessity. Similarly the slowness and imprecision disappear once the timescale is understood. Deep time allows millions upon millions of years for wings to grow, eyes to evolve, thumbs to twist around the hand to oppose the fingers and even for a primate brain to attain self-awareness. Every serious attempt to discredit Darwinism founders before the apparently miraculous workings of deep time. Once organic molecules have formed by whatever process, then, assuming roughly stable conditions, camels, elephants and men are inevitable. By a curious inversion of thought that emphasizes the extraordinary and unique nature of Darwinism as a theory, Aristotelian final causes had, in one peculiar sense, been revived. Man *was* present in the amoeba.

But, with or without traces of Aristotelianism, of all science's arrogant assaults upon the religious realm, this was the most appalling and fundamental. As the biologist Richard Dawkins has written, 'Darwin made it possible to be an intellectually fulfilled atheist'.[17]

Before Darwin the argument from design sustained God as the organizer of complexity. After Darwin the greatest complexity of all – organic complexity – was utterly accounted for by the workings of deep time. Worse, we were no longer the children of God, we were the descendants of apes and algae. All our former pride in ourselves as Lords of Creation melted away to be replaced by the awareness of our dumb animal blood. Science's statements of the futile contingency of our existence must now have gone as far as it dared.

'In endless space,' wrote Arthur Schopenhauer, the nineteenth-century philosopher of heroic pessimism, 'countless luminous spheres, round each of which some dozen smaller spheres revolve, hot at the core and covered over with a hard, cold crust; on this crust a mouldy film had produced living, knowing beings . . .'[18]

'I feel', wrote Darwin about the problem of morality in the world which he discovered he had created, 'most deeply that the whole subject is too profound for the human intellect. A dog might as well speculate on the mind of Newton. Let each man hope and believe what he can.'

In such words Darwin emerges as the very modern figure of the scientific tragedian. This is the world, he is saying, whether we like it or not; we may not want it to be so, but it is so. All else is wishful thinking. Belief cannot be separated from this truth. It must reconstruct itself in the cold-eyed awareness that the facts of the world can be of no help. It was the same as the message we had derived from Newton's cosmos of absolute time and space, but the simplicity of Darwinian natural selection was an unprecedentedly brutal form of that message. We were accidental animals.

In our century this fatal and humiliating connection to nature was further celebrated by the discovery of the molecular structure of deoxyribonucleic acid (DNA). DNA is the carrier of the genetic message. In some reductive sense it might be said to contain all that we, and most other organisms, are. Blue eyes, brown hair, short, tall, thin, fat are all said to be encoded in this fabulously complex molecule. But, though the analysis of the structure of DNA by the Anglo-American team of J. D. Watson and F. H. C. Crick did reveal complexity, it also revealed an appallingly finite simplicity. The twin spiral of nucleotide chains with their connecting systems of four bases provided a model that was at once visually understandable. Indeed, it was possessed of a kind of poignant beauty. The imagination seized on the image – reproduced hundreds of times in film animations – of the unfolding spirals as the delicate, intricate thread of life which joined us to each other, to our ancestors and to all creation.

Near the climax of the film *ET* when the lovable alien is close to death in spite of all the efforts of contemporary technology, a scientist rushes in through plastic curtains, the drips and the monitoring machines to announce that the alien has DNA. Amidst the more tumultuous emotion of the moment, the remark is almost lost. But it was a way of saying that the alien was like us, a cosmically reassuring statement. It was as if the film-makers' imaginations were possessed of the final, arrogant sentimentality – that nothing could be utterly strange, that somehow our lives, reduced as they might have been by all that we knew, were patterns of all life. But the implicit celebration is unconvincing because DNA, for all its evocative beauty, remains a chemical machine.

The understanding of DNA has now led to an international scientific project to map the human genome – all the genes that we possess. It is a simple but enormous project costing billions. When completed, it will provide us with a map of the disposition of all the instructions and codes that make us what we are. Few doubt that it is possible, but many will wonder about the meaning of the phrase 'what we are'. For all the sentiment and beauty that attends the twin unfolding spiral of DNA, the message of the image is as bleak and unforgiving as Darwin's. But the difference between the twentieth century and the nineteenth can be glimpsed in the fact that neither Watson nor Crick felt Darwin's heroic sense that something unfaceable was being faced; they felt merely that the model worked by being in accord with the known facts. For them, it was enough.

The concluding phase of this process of the humbling of man is, however, curiously ambiguous. Many will say that the final nail driven into the coffin of man's self-esteem was not driven by science at all, but by a kind of art. Yet the impulse behind the hammer of psychoanalysis was scientific. Its creator, Sigmund Freud, would grow irate at any suggestion that what he was doing was anything other than the purest, hardest science. When Havelock Ellis described him as an artist rather than a scientist, he described this attitude as 'the most refined and amiable form of resistance, calling me a great artist in order to injure the validity of our scientific claims.'[19] Resistance to the scientific truth becomes, under the gaze of Freud, a form of disorder.

There were many cultural reasons for Freud's insistence on his role as a scientist. The late-nineteenth-century Vienna from which he sprang viewed science unquestioningly as the most powerful antidote to the ills of the world and Freud's own Cartesian need for clear knowledge provided obvious psychological pressure. 'Scientific knowledge', he said, 'is the only road which can lead us to a knowledge of a reality outside ourselves.'[20]

But there was also the pressure of the logic of the history of science. This pressure was inward. By the end of the nineteenth century it was quite common for scientists to believe that all the major questions of physics were close to solution. 'It only remained', Sir William Dampier assured the world, 'to carry measurements to the higher degree of accuracy represented by another decimal place, and to frame some reasonably credible theory of the structure of the luminiferous ether.'[21]

Such views were to prove spectacularly wrong. But the point was that, to the scientifically enlightened, substantial knowledge of the universe and the nature of matter was in place. Now, after Darwin, life itself was beginning to succumb. Logically the human mind, the generator of all this reason and knowledge, would be the next frontier to be crossed.

Freud could thus see himself quite clearly as the inheritor of the scientific impulse of Newton and Darwin. And he did so, adding a characteristically pessimistic twist to the idea. First Copernicus had turned us into a cosmic speck, secondly Darwin had robbed us of any privileged position in creation and, finally, he, Freud, had shown that man was not even master of his own mind. *But* the scientific procedure offered the possibility of a consoling mastery.

'Man's observation of the great astronomical regularities,' Freud wrote, 'not only furnished him with a model for introducing order into his life, but gave him the first point of departure for doing so.'[22]

This was the unmistakable sound of a great human project coming home to roost. In that remark Freud, in all his disciplined intensity, was attempting to close a vast historical circle. Science may have told us we were lonely accidents, but it did, at least, provide a model of order and harmony which we might apply to our own lives. Value was to be rediscovered in the application of the ordering mastery of the scientific discipline to our own inner selves. We must turn away from the inherent tragedy of the vision and get down to work.

Freud's own project was the creation of a narrative of ourselves. He pursued the scientific search for causality and origins into the life of the individual. His most recognizable contribution to the imagination of the world is the elaborate bridge he constructed between the adult and the drama of childhood. Beginning by studying pathological conditions, he moved on to the universal drama of the self as a series of conflicts through which we were obliged to pass to attain any degree of stable maturity. The human self became, in Freud's hands, a stratified system of id, ego and superego. Our very consciousness itself was but a thin layer upon the raging storms of the subconscious to which we could find access via analysis and dreams. Driven by an erotic and, he later concluded, a death instinct, we lived our lives in a state of dynamic compromise. Civilization, the world and our own bodies demanded a renunciation or thwarting of the instincts and, therefore, a permanent state of radical discontent.

Freud identified our triple misfortune: 'the superior power of nature, the feebleness of our own bodies and the inadequacy of the regulations which adjust the mutual relationships of human beings in the family, the state and society.'[23]

And, in a precise echo of the resigned words of Darwin, he surveyed the significance of such a quandary for the destiny of man: 'Happiness in the reduced sense in which we recognize it as possible, is a problem of the economics of the individual's libido. There is no golden rule which applies to everyone: every man must find out for himself in what particular fashion he can be saved.'[24]

The terms of Freud's narrative of the self and his methods have often been questioned both as to their scientific validity and as to their truth. There are many rival psychoanalytic systems as well as profound modifications of Freud's own views among Freudians. Scientifically, his status is routinely questioned because of the curiously unchallengeable way in which his theories are stated. Nothing, it seems, can disprove the Freudian narrative since it does not make precise predictions in individual cases, nor can it be statistically tested. We can test Newton by observing the motion of the planets, Darwin by observation of organic species and their environments, but we cannot test Freud by observation of any individual or collection of individuals. The theory exists and all variations can be adapted to its demands.

Yet, as an idea, psychoanalysis has been fabulously successful. It has become, in our age, a pervasive orthodoxy of self-knowledge. The subconscious is embraced as excuse, explanation and domestic curiosity. We now conceive of ourselves as, above all, stories with interwoven themes and hidden symbols, motifs and significance. It has provided us with an internalized validation of our demand for explanations and origins. It both links us to our neighbours and separates us from them.

In this context as well as in the context of Freud's imagination, the issue of whether or not his system is classifiable as scientific is beside the point. What matters is that he conceived of his work as the climax of a scientific tradition of sceptical inquiry which, 250 years after Descartes's Cogito, had finally turned inward to confront the true nature of the self upon which the great sceptic had once based his all-conquering epistemology. The philosophical problem of knowledge is reduced to the problem of how to subsume all knowledge into the empire of science.

Freud's vision was the climax of a project that had conclusively humbled our religious arrogance and scorned our humanistic hubris. Understanding the present means, to a great extent, grasping something of his conclusive, tragic vision of man. His work may not be science, but there can be no doubt that his tragic, literary grandeur and imaginative power finally deliver the one clear message that science has wished to pass on to us ever since Galileo applied his eye to the telescope: that we are nothing but trivial accidents and that each man must hope and believe what he can in the grim certainty that nobody and nothing will ever be able to tell him whether he is right or wrong.

4
Defending the faith

'Ah, love, let us be true
To one another!

Arnold[1]

The story I have been telling is a simple one. It is, in a slightly more elaborate form, the same story as the primitive tribe introduced to penicillin. It is the story of a culture – our culture – being progressively overwhelmed and transformed by science.

All the subtleties of Descartes, Hume and Kant are, at heart, merely different ways of confronting this invasion. Their solutions are complex, refined and difficult because they have to be. The challenge of science is a challenge to all that we are and all that we know. The response to such a challenge might *finally* be very simple. But the difficult process of developing such a solution cannot avoid surveying and analysing the full complexity of human life.

The official, popularizing versions of the story, written by scientists or philosophers like Bertrand Russell who allowed themselves to be seduced by science, are very different from mine. For them the story is heroic, a great human struggle to free ourselves of the shackles of old illusions and confront the one certain truth that it is our particular destiny to grasp. Science, in this view, is a triumphant human progress towards real knowledge of the real world. This is the official view of the schoolroom and the television spectacular. For me it is nonsensical propaganda which conceals all the important issues. In my version the story is a sad one, a long tale of decline and defeat, of a struggle to hold back the cruel pessimism of science.

The key to this struggle, it cannot be said too often, is the way in which science forces us to separate our values from our knowledge of the world. Thanks to Newton we cannot discover goodness in the mechanics of the heavens, thanks to Darwin we cannot find it in the phenomenon of life and thanks to Freud we cannot find it in ourselves. The struggle is to find a new basis for goodness, purpose and meaning.

The pain of this separation of knowledge and value can be understood through the different ways in which we contemplate nature, because it is from nature that we long for the reassurance that we cannot have. In an essay entitled *Nonmoral Nature*, the contemporary American science writer palaeontologist and biologist Stephen Jay Gould, an eloquent defender of the hard truth of hard science, discusses the strange case of the ichneumon wasp. During its larval stage this creature lives as a parasite, feeding on the bodies of, usually, caterpillars. The female adult injects her eggs into the host and victim via a long thin tube known as an ovipositor. Some varieties of ichneumon lay the eggs on the surface, so, as a precaution against them being dislodged, they simultaneously inject a paralysing toxin to prevent the host from moving during the process of incubating and then feeding their offspring. To keep the food fresh, this toxin paralyses but does not kill. For the same reason larvae deposited inside the caterpillar follow a particular eating pattern designed to consume inner organs and tissue in such a way that the host will continue to live for as long as the larvae require.

Gould points out how the life of the ichneumons captured the moral crisis of the nineteenth century. The very exploitative viciousness, the cruel calculation of the wasps' behaviour seemed to deny the possibility of a benevolent universe. It was one thing to eat your prey, quite another to contrive to keep it alive while you did so. The Victorians attempted to be objective about this terrible spectacle. They made serious attempts not to see nature in terms of human morality. They wished to distance the horror by scientific objectivity. But, as Gould points out, they found themselves obliged to employ the language of human drama simply to tell the story. Our words are loaded with values and they seem able to trap us into involvement with the fate of the caterpillar.

'We seem', comments Gould, 'to be caught in the mythic

structures of our own cultural sagas, quite unable, even in our basic descriptions, to use any other language than the metaphors of battle and conquest. We cannot render this corner of natural history as anything but a story, combining the themes of grim horror and fascination and usually ending not so much with pity for the caterpillar as with admiration for the efficiency of the ichneumon.'[2]

A hundred years later our language is closer to final defeat by science. We do not pine for goodness in nature as passionately as the Victorians. We have found ways of softening its terrors.

I recently visited the Natural History Museum in London. Completed in 1881, it is a building that stands as an emblem of the high Victorian belief in a scientific understanding of the world. Constructed in a flamboyant Romanesque style with a vast, barrel-vaulted interior space, now occupied by a dinosaur skeleton, it asserts the continuity of the scientific culture. It is a monument to the English legacy of Francis Bacon, a storehouse of the data that will underpin inductive truth.

Now, of course, the confidence that inspired this building has been lost. The museum has been modernized. Mute, stuffed beasts were once enough: their irreducible presence among so many thousands of others was sufficient wonder for the Victorian sensibility. But all that is slowly giving way to hotter, sweeter thrills. Now there are complex, interactive displays designed to teach the basics of biology and zoology to children and impatient, uncultivated adults. Buttons can be pressed, screens watched and models manipulated. Amidst this carnival of clutter and diversity, one noisy, colourful exhibit is called 'Creepy-Crawlies' and there I found a giant model of the female ichneumon wasp frozen in the act of injecting her eggs into a caterpillar.

A Victorian horror story has become a modern celebration of intriguing diversity. Do not feel sorry for the caterpillar, the model seems to be telling us, applaud the wasp for its ingenuity. It is no good weeping human tears over inhuman nature.

But what about faith? How did religion itself cope with this terrible onslaught, this appalling dislocation?

Perhaps the questions were not worth asking. Perhaps we should confront the faithless universe with a new heroism. That was the attitude of Friedrich Nietzsche (1844–1900). He contemplated the refinements of the great Enlightenment philosophers' attempts to forge a new definition of truth and value and a new defence of religion. He lost his temper. He called Kant 'a catastrophic spider'. The Königsberg ascetic had woven his metaphysic out of the Enlightenment's epistemological crisis and trapped us all like flies. Nietzsche regarded the entire effort with grandiose disgust, calling both Leibniz – the prophet of 'pre-established harmony' – and Kant – the supreme defender of the moral nature of man – the 'two greatest impediments to the intellectual integrity of Europe.'³

All their metaphysics, he thought, were no more than a craven attempt to save mankind's cowardly humility and its God. But, in the face of the colossal structure of our own knowledge, we did not need some crabbed shuffling of the theological pack. God was dead. But our new knowledge revealed not that we were impotent, but that we could become gods in his place. It would, Nietzsche thought, take us two centuries to face this transformation in all its aspects. But, once we had faced it, we would be free. The long birth of this new age, however, would result in unprecedented strife. Nietzsche's own work signalled the onset of labour.

'There will be wars,' he wrote, 'such as there have never yet been on earth. Only after me will there be *grand politics* on earth.'⁴

What was disgusting to Nietzsche's ambitious nineteenth-century mind was the attempt to preserve the Christian fabric against the onslaught of Enlightenment knowledge. The idealism which had allowed Kant to slip the bonds of material reality was an undignified retreat of the European soul. The effort appeared cowardly, dishonest, deluded. Even Luther was condemned. His rebellion was no more than a feeble attempt to save rather than overthrow the Church. Protestantism and idealism were no more than absurd and contrived defensive systems.

'The *lie* of the ideal', Nietzsche wrote, 'has hitherto been the curse on reality; through it mankind itself has become mendacious and false down to its deepest instincts – to the point of worshipping the *inverse* values to those which alone could guarantee it prosperity, future, the exalted *right* to a future.'⁵

Our real destiny ought to be the cold, heroic confrontation

with truth – 'Philosophy, as I have hitherto understood and lived it, is a voluntary living in ice and high mountains . . .'[6] We were to accept the role implicit in the genius of Newton. We were to become gods, self-creating and self-defining, free at last from the cnoking mythologies of the past.

This was the heroic, individualistic response to the imaginative crisis inspired by the scientific project. It represented an attempt to create a cruel, hard, aristocratic religion out of atheism and the lonely truth. New values would be heroically forged by great souls. This was all that ultimately mattered. Not all men were irreducible ends in themselves, as Kant had dreamed, only the chosen few.

The Nietzschean solution was, in effect, to start again now that the values and mythologies of the past had been so thoroughly discredited. It was an influential response that was to wash ashore in our own century in any number of disguises. Today Nietzsche has been both liberalized and turned into the precursor of Nazism. Neither is quite fair. His role was simply to see the problem with such tortured clarity that it could never again be ignored. In his final years he descended into insanity.

But, for most thinkers, starting again represented a kind of defeat. It meant throwing away the whole history of religious insight and truth. Perhaps the more sober, saner response was to find new ways of defending the ancient faith. The strength of this approach lies in the obvious human inadequacy of science. On the one hand it had destroyed religion's foundations; yet, on the other, it refused to provide the kind of answers religion could offer. We could have the truth or we could have a place in the world but we could not have both.

Science was the lethally dispassionate search for truth in the world whatever its meaning might be; religion was the passionate search for meaning whatever the truth might be. Science can lay a claim to a meaning in the sense of establishing causality, and religion could claim truth in the sense of a transcendent order. But science's meaning does not answer the question Why? And religion's truth had no scientific relevance.

Above all, the division between truth and meaning persists, for those are the way the terms are defined in the modern world: truth and meaning were severed by knowledge. That is what we think we know. Draw no conclusions from the private life of the ichneumon wasp, just celebrate that fact that we know about it.

The difficulty of this position produced, in the early nineteenth century, an intense, romantic suffering. In 1819 the English romantic poet John Keats wrote: ' "Beauty is truth, truth beauty," – that is all/Ye know on earth, and all ye need to know.'[7] He wrote it precisely because it was not true in the world of the early nineteenth century. It was rhetoric or wishful thinking. And, in Keats's use of the word 'all', we can feel the pressure of the romantic revolt against the cosmic hubris of scientific man. We did not *need* his truths, we did not need more than the synonimity of beauty and truth. To say that his motto was all that we needed to know was to demand a new innocence, a rest from the turmoil of our knowledge. It did not, however, offer a programme of action other than perpetual aesthetic languor.

Yet a programme was required if anything, other than science, was to survive.

As if seeing into the future, Kant had defined a way of defending God against Darwin and Freud. He had seen the dangers of attempting to carve out some specialized niche for him amid the truths of this world – science would only come along and mock those as it had mocked the physics of transubstantiation and the cosmologies of geocentrism. To define God in the world was to condemn him to a permanent retreat in the face of the rigour of scientific analysis. The world of the senses was at the mercy of science.

But, in expelling him from the world of the senses, Kant had created a figure far removed from the immanent and effective God of the Middle Ages. Instead of the master of the benign fabric which placed a farmworker in the stained glass of a cathedral whose totality was an architectural vision of the intellectual unity of creation, there was the infinitely more subtle revelation of God in the deep structure of the human soul.

The question was – and is – whether religion can survive on the basis of such subtlety, whether a material world without miracles or meaning can still sustain the faith. And, in turn, that question is dependent on what we mean by religion and what constitutes faith.

First, it is clear that there is something about the human

condition that demands a dimension we call religious, whatever it might be. Particular faiths have come and gone, but nothing has ever displaced the religious presence itself from human life. It has always accompanied men and their cultures.

Religions have usually attempted to relate their spiritual systems to the material experience of the world. In doing so they have depended on the conviction that value and meaning can be found in the facts of the world – precisely the conviction that science has so successfully defied and apparently disproved. It is, therefore, idle to pretend, as many do, that there is no contradiction between religion and science. Science contradicts religion as surely as Judaism contradicts Islam – they are absolutely and irresolvably conflicting views. Unless, that is, science is obliged to change its fundamental nature.

In early societies the cycles of agriculture produced transcendent explanations of the changes of the seasons or, most commonly, worship of the sun. Science may tell modern man that such repetitive rhythms are all but immune to immediate failure; the sun will rise, the seasons will pass because of the relative stability of the solar system as a whole. But, once, they appeared as precarious as the life of man himself. So precious and so mysterious were the rhythms of life that their continuance was a matter of daily anxiety. In the case of the Aztecs, for example, the sun-god was obliged to defend himself against enemies every night, so the coming of the dawn was a constant uncertainty and its daily appearance a military triumph against unspeakable odds.

But religions which still clung tightly to such natural cycles could also be seen to be tied closely to the particular human societies from which they sprang. They were local, specific faiths. They did not aspire to be Theories of Everything.

A change began in 1200 BC when Moses formalized Judaic theology. This was the first of a number of new and more inclusive systems that were to spring up around the world. Unlike their forerunners, these were the beliefs of sophisticated people who could remove themselves for a time from the urgent and exclusive demands of agriculture. There was a surplus of intellectual energy available to contemplate the whole of life. The new systems had in common a complete explanation of all human life and history and, above all, they were rational.

'The process of rationalization', Max Weber wrote, 'favoured the primacy of universal gods; and every consistent crystallization of the pantheon followed systematic rational principles to some degree, since it was always influenced by professional sacerdotal rationalism or by the rational striving for order on the part of secular individuals. Above all, it is the aforementioned relationship of the rational regularity of the stars in their heavenly courses as regulated by divine order, to the inviolable sacred social order in terrestrial affairs, that makes the universal gods the responsible guardians of both these phenomena.'[8]

In other words: religion, like science, began with the inscrutable and majestic spectacle of the heavens. This points again to the fact that they are destined to compete: they are occupying the same territory.

The great religions, therefore, were about completeness, a totality of explanation. After Moses, in 1000 BC the Rig-Veda was written down in India and was followed, in 600 BC, by the Upanishads. Siddhārtha, the Buddha, taught around 500 BC. Zoroastrianism began in Persia in 660 BC. Confucius was born in 551 BC and so on. For the 1800 years up to the death of Muhammad in AD 632, the world seemed to have embarked upon a massively diverse programme of universal explanation. And, for such explanations to be true, they had to apply to all aspects of life. Religion progressed from its roots in the cycles of nature and as a background to culture to become the culture itself. In Chinese, Indian and European civilizations, religion aspired successfully to become one with all the works and lives of men. In Christian Europe the grandest expressions of this unity were the Gothic cathedrals.

The explanations and justifications in each of these systems were, of course, extraordinarily diverse. Weber characterized each by the ideally perfect carrier of each faith: 'In Confucianism, the world-organizing bureaucrat; in Hinduism, the world-ordering magician; in Buddhism, the mendicant monk wandering through the world; in Islam, the warrior seeking to conquer the world; in Judaism, the wandering trader; and in Christianity, the itinerant journeyman.'[9]

But they were all explanations and justifications of human life and all tended to fall into the prophet-priest pattern also described by Weber. Prophets provided the system and the ultimate values; priests analysed and rationalized this system and adapted it to the forms and customs of life. It is an important pattern in human affairs which was to be repeated in the development of science. The prophets were the innovative scientists, the priests were the interpreters, extenders and technologists who followed in their wake.

Yet from one of these Theories of Everything – only one – sprang the form of knowledge that was to challenge and transform them all. There are any number of theories as to why the scientific imagination should have sprung solely from the Christian. None is conclusive, but some points are worth making.

First, of all the universal religions, Christianity was, perhaps, the most radical. Like the Pythagorean community at Croton in ancient Greece which had worshipped the purity of number, the Christians considered the body as a prison and viewed life on earth as a preparation for Heaven. In spite of the efforts of the Middle Ages to unite theology with Aristotle, this was, at heart, a Platonic vision that specifically dismissed appearances in favour of essences. And, as I have said, the *Scienza Nuova* was closely linked to a Platonic revival. Platonism, science and Christianity all shared the conviction that there was an underlying order behind the accidents of this world.

In Christianity this wisdom became symbolic. Theologians interpreted the life of Christ as replete with significance. From the centrality of the bread and wine at the Last Supper to the details of his nativity at Bethlehem, all could be minutely meditated upon for wider and deeper meanings. The facts of the world were symbolically linked to a divine order and were, therefore, directly imbued with meaning and value. In the years of the Catholic Church's decadence so-called fragments of the True Cross were sold across Europe as if all matter that had played a part in that drama had been magically transformed. Yet even this extremity of superstition can be seen as a precursor of the new age – the scientific imagination was also founded upon an obsessive observation of specific detail and upon the deeper significance of matter. Christianity established the style of the new knowledge. It was just that science did not save your soul.

Furthermore, Christianity was an individualizing faith. In the

Church itself this emphasis has ebbed and flowed. But the inescapable centre of the Christian doctrine is the suffering and spiritual progress of one man, Jesus, who was humbly born into the daily rituals of humanity.

There is nothing in any other faith to compare with the figure of Christ on the cross as an emblem of the trials of human life. It speaks simultaneously of the reality and complexity of the things of this world as well as the profound humanity and loneliness of the effort required to attain the next. The attempt of the medieval Church to contain this dynamic humanity in Christianity within a static, Aristotelian/Thomist universe can be seen in this context to be a kind of betrayal. The Franciscan side of Christianity had been right to see potential sin in subtlety and the pride of the intellect. Thomism was *la trahison des clercs* – the heresy of the over-refined.

Yet it was a heresy that was unavoidably built into the individualistic structure of Christianity. And it was this intellectual individualism towards which the faith had evolved in the hands of Aquinas that permitted the emergence of science. The scientists were the new suffering Christs rebelling against the suppression of their new forms of knowledge by a decadent ecclesiastical authority.

But perhaps Christianity's most powerful claim to be the sole creator of the modern world derives from its underlying tragic sense. The world destroyed its saviour. God sent his Son to become human and to suffer and die as a human. The orthodox Christian would say that process was an exemplary identification of the divine with the human. The danger is that the drama could become all too human. The suffering and death could still have meaning without an external creator. Perhaps, in becoming flesh, God died. Perhaps the story tells us that the truth is here, now and within, rather than in some distant paradise. And, if that is so, perhaps it is here, now and within Einstein, Newton or Galileo as much as in Jesus or St Paul.

These are generalities. More puzzling is why science did not emerge in the highly developed civilizations of the East. A number of ingenious reasons have been suggested: a rigid social structure preserving learning within a literary ruling class, written language remaining aristocratically distant from technical and everyday language, contempt for manual labour holding back technology, the size of the Chinese empire and so on. It is a vivid enough contrast that summarizes all these points to hold up the ideal of the Confucian scholar against the figure of Newton. The Confucian was a patrician who, according to tradition, would grow one fingernail to enormous length to demonstrate how far he was above lowly manual work. Newton ground his own lenses. To become a scientist the patrician would have had to break his nail.

Yet, perhaps more important than any of this for the purposes of understanding the present, is the fundamental intellectual difference that lay behind all these details. Chinese religion was holistic. True knowledge was knowledge of the whole, not the parts. Experiment in the Galilean or Newtonian sense would thus be meaningless. The quasi-ideal conditions of a frictionless surface, of weights dropped from a tower or of light split by a prism would be trivial to a Chinese scholar next to an understanding of the greater harmonies of nature and society. And so they were, but triviality was to mount upon triviality until universal laws of unprecedented effectiveness were revealed.

As John Barrow has written: 'Ancient holistic ideas provided no methodology for developing understanding, because they outlawed the concept of cutting up nature into manageable independent pieces that could be understood individually.'[10]

This is an important distinction which, in fact, can be discovered in comparisons between Christianity and a number of other religions. Christian emphasis on the details of the life of Christ inspired a cultural acceptance of the study of parts, of a fragmented expertise. To some Oriental faiths an understanding of parts was no understanding at all. It was self-evident that all things were one. In such a context successful science could not even begin.

Perhaps, finally, monotheism itself was the ideal environment for science. A single, all-powerful God would encourage the view that uniform laws lay hidden beneath the surface of nature. And it is significant that the 'mind of God' is frequently evoked by scientists as a more poetic version of what they are examining than the more usual 'reality'. This is not done simply to acquire the virtue of God, it is also because God as an individual seems to conform with the scientific faith in simplification. Newton himself was a secret unitarian – he did not believe in the orthodox Christian concept of the Trinity – and it is clear that the belief directly linked to his drive towards a perfect, unified synthesis of scientific knowledge. Openly confessed unitarianism, however, would have excluded him from the professorship at Cambridge.

For whatever reason, Christian Europe created science, the devastatingly effective system that was to call into question all the world religions that had been codified in the previous 3,000 years. So it was the Christians who had to endure the first imaginative onslaught.

We have seen the high intellectual repositioning that took place from Descartes, through Hume to Kant. But this was the most rarefied level of debate, concerned primarily with the urgent but practically remote problem of what and how we could know and almost avoiding the desperate struggle of faith itself. At the more immediate level of how society could work in the light of the new knowledge there seemed to be more obviously pressing issues than epistemology.

The medieval synthesis was based upon the belief that all was religious. Society flowed smoothly downward from kings and popes to peasants. There were abuses, protests, dissent and doubt, but the underlying model was clear. The turbulence – intellectual and political – of the sixteenth and seventeenth centuries, however, was more than local dissent. The underlying model was being challenged. It is almost banal to point to the hundreds of references in the works of Shakespeare to the appalling dangers of a disordered state, of the destruction of hierarchies. But they are there, and Shakespeare was born in the same year as Galileo.

The Galilean message, his way of creating a co-existence of science and religion, was that there were two books – the book of faith and the book of nature – rather than one universal book, one *Summa*, that made sense of all things political, moral and cosmological. And, if there were two books, then in any area of life we would have to consider which one to consult. Indeed, it could be the case that our public lives may be conducted entirely according to the book of nature, leaving the book of faith only for our private, inward journeys. The unity of the religious world was thus undermined by the explicit acceptance that public and private morality could be reasonably separated. What we say – on the basis of the book of faith – is not necessarily what we do – on the basis of the book of nature.

The effect of science with its individualism and its insistence on observation and reason as opposed to authority is clear enough. It was a condensation of the tendencies of the age as well as their most effective expression. Like the voyages of discovery, the rapid mercantile growth of Europe and the Protestant questioning of the nature of Christianity, it represented a dynamic and progressive view of human life. It speculated, debated and experimented. As Descartes's method had established, its ideal was sceptical and questioning and its only standard was the interior of the questioning mind.

Furthermore, it appeared limitless. If the Book of Faith were to be separated out, then science had the whole of material creation as its plaything. So, for example, human society might be capable of scientific analysis. And, if it were, there would be no need to persist with the religious politics that had plunged Europe into war. Science could provide a rational model.

But the abyss that lies between any such model and a religious society is immense. In a scientific society, reason would have to prevail. There could be no subjection or oppression of reasoned analysis in the name of any extra-rational authority. Equally, a scientific society would, in the long term, be classless. Each man's conscience was his own as was his reason. The next Newton could come from any stratum of society.

Such considerations would slowly penetrate European thought and form her societies in the years of scientific progress. They were accompanied and echoed by successive attempts either to halt or to collude with the assault on religion. The response of a straightforward, fundamentalist denial of science's insights persisted and is with us still. Yet even the Catholic Church abandoned this defence. Having fought back against the new philosophers – scientists – by comparing them to the builders of the Tower of Babel who wished to scale the heavens and rebel against God, they finally came to terms. The Jesuit deal implicit in the Counter-Reformation was that if the individual of the new age would surrender his moral autonomy to the Church, then, in return, the Church would relax the more severe and ascetic demands of medieval religion. And, in 1893, Pope Leo XIII's encyclical *Providentissimus Deus* officially endorsed the Galilean view of relations between scientific and biblical truth.

But it was Protestantism that was to provide the most dynamic image of faith's struggle against the inroads of science. For a start the Reformation had been born out of the same turmoil and the same imaginative changes as science. It had also been inspired to reject authority – that of the Catholic Church – in the same way that the first scientists had rejected authority – that of the Church and of classical learning – as a generalized guide to the understanding of the universe. Crucially Protestantism also emphasized the centrality of reason and of language as opposed to the unquestioning acceptance and grand mystifications of the medieval Church. In Protestantism the magical element of religion was at first played down and then, at last, firmly suppressed.

'In Christianity,' Max Weber wrote, 'the importance of preaching has been proportional to the elimination of the more magical and sacramental components of the religion. Consequently, preaching achieves the greatest significance in Protestantism, in which the concept of the priest has been supplanted altogether by that of the preacher.'[11]

Abandoning magic, it seems to me, is crucial. The power of magic is popular belief. It can and must 'work' only within the context of such belief. If everybody believes in encounters with demons, then the encounters occur. A private, personal magic is impossible, it must exist within a culture. Modern scepticism about the 'reality' of such encounters is beside the point. But this is precisely the area that science most effectively invades. By its displays of predictive power – as in the case of the return of Halley's Comet – it draws belief away from magic, its deadly rival. Perhaps only the words are changed: magic becomes science, magicians become scientists. But the change still occurs.

In reducing the magical aspect of the faith, Protestantism must have improved its strategic position. It had simply abandoned territory that could not be defended. This, combined with the decisive Protestant emphasis on the struggle of the individual soul, opened up the possibility of radical new definitions of religion. It may have taken the Catholic Church until 1893 officially to acknowledge that it may have been wrong in the case of Galileo, but, by then, Protestant thought had already recreated the faith.

Kant and Hume were the great initiators of this Protestant enterprise. They were to be followed by a decisive phase which still dominates most theological thought today. This was the development of 'liberal' theology.

Liberalism in theology springs from Hegel (1770–1831) and from the desire to unify the whole world picture, including science, into a religious system which could not simply be falsified. The Hegelian vision was of history as the unfolding story of a single spiritual development. The point was the unfolding. This allowed for progress and change instead of insisting on the unity of a single, revealed truth. Science could thus be embraced as a part of the faith. The knowledge provided by science was as much part of this process as anything else and in no way invalidated its religious truth. Science was simply a further phase in the revelation of the great historical system. Truth was an unfolding, a forward movement towards some ultimate condition, traditionally known by Christians as the Kingdom of God.

The Hegelian goal was human freedom, but from this man was held back by necessity and alienation. Necessity was his dependence on nature to feed, clothe and sustain himself. This science could help him overcome. But his path to freedom and spirituality was still blocked by his alienation. Man saw himself as subject and object. He saw himself in the world but also as other, as somehow separate. 'Alienation' is the Hegelian form of the problem of scale identified by Pascal, Lewis Carroll, Jonathan Swift and countless others. It is the perpetual ambiguity and puzzle of consciousness.

From such a position two developments are possible. The first is to abandon the religious interpretation completely. Indeed, Hegel's disciple Ludwig Feuerbach specifically located the source of alienation as religion itself. Feuerbach called himself Luther II and insisted that man would only be free if he finally demythologized religion and placed himself, rather than God, at the centre of consciousness. The great narrative of historical development thus becomes a purely human story.

The point on which Feuerbach had seized was that the imaginative power of the central Hegelian view of history as an unfolding story with distinct and identifiable processes at work was such that the religious backdrop was hardly necessary. This is a familiar insight. We have seen how Newtonianism could survive as physics stripped of this God and his magic. Similarly Descartes's God was insufficiently glued on to his scepticism to endure. Always the tendency of our age is the same: to take only what we think we need from the past and leave behind that in which we can no longer believe. We edit the culture until it accords with our own image of ourselves.

Perhaps the supreme act of editing of Hegel was Marxism. Karl Marx simply replaced the religious determinant of the pattern of history with economic and social structures. Here alienation was located in the workplace where modern man was condemned to be the tool of capitalist processes with no interest in or identification with what he produced.

Marx represents the highest point of the attempt to link politics and science. Like Nietzsche, he was not content with the thoughtful impotence to which philosophy had been reduced. 'The philosophers', he wrote, 'have only interpreted the world in various ways; the point is to change it.'[12] This is not a mere rhetorical flourish, for it represents a fundamental inversion of the conventional thought process. Conventionally we might imagine that people would think about the world and then attempt to change it into something more in line with their conclusions. But Marx saw that the very thought processes were determined by the material reality of the society that produced them. It was pointless interpreting the world when your interpretations could be no more than expressions of that world. And any such expression could be no more than part of the process of interminable philosophical conflict that lay beneath the capitalist world. The resolution of all such conflicts – 'contradictions' they are usually called in the jargon – could only be achieved in a communist world, so social action aimed at hastening the arrival of communism must take precedence over interpretation. The paralysis of alienation was thus circumvented by action that pre-empted thought.

This places the emphasis on the effectiveness of the Marxist belief in Marxism itself. You had to have faith to act without thought. But this faith needed more if it were not to descend into mere Utopianism, if it were not too obviously to be mere faith. The extra was provided by Marx's conviction that he had discovered the scientific laws of social change – primarily the historic movement from primitive socialism to feudalism, capital-ism and, finally, true communism. This was a 'scientific' fact which did not require individual intervention or commitment. Indeed, there was no ethical element whatsoever. Once the facts of social change were made clear to the proletariat, the class that would force the next phase of change, then the revolution would take place. This would then produce the change in consciousness that would effect the necessary moral transformation and true communism would ensue.

Marx's science was the economic evolution of society. Discern-ing Hegelian patterns in history, he used these to produce fore-casts. He created a powerful, deterministic, atheistic system as the full and final explanation of human history. The scientific God of Causality was shown to apply to social and political structures.

Of course, the thought assumed, and still does, far more than we could possibly know. Nothing that can comfortably be called science has yet emerged from economics, politics or sociology. In his eagerness to borrow the imaginative, persuasive power of science, Marx had produced a strange distortion of history. He assumed, for example, that economic growth was a permanent feature of human society and that the rapidly industrializing world that he saw about him was a definite product of a single, linear, historical narrative. But the growth and the progress which formed such a central part of his 'science' were, as I have said, only recent developments. Societies have existed in conditions of economic stagnation far more often than they have enjoyed economic growth and the Marxist phases of history are an appallingly crude generalization.

As a result, of course, Marx's forecasts were uniformly wrong. Applied to politics in the real world, his thought produced mon-strous, destructive regimes. Yet the impulse behind Marxism is understandable enough – surely human reason, having reached so far, could reach into the workings of human society. If we could understand and forecast the activities of the heavens, our own world and its history could scarcely be beyond our grasp. Furthermore, Marxism possessed an irrational, imaginative power thanks to its obvious indebtedness to the Christianity it was designed to overthrow. In Christianity there was a fallen state, followed by revelation of the coming of Christ and, finally, there was salvation. In Marxism there was the fallen state of bourgeois capitalism, the coming of the Revolution and, finally, the onset of communism. Both processes required a radical transformation of the human spirit.

But, for my purposes, the important question about the Marxist effort was not so much why or how it was wrong, but, rather, whether it could possibly have been right. Is a rational, scientific, conclusive and universal understanding of human history possible? Marx could have been wrong because his observation was wrong, because his analysis was wrong or because he did not understand practical politics. But was he wrong because there was no possibility of being right? Does history lie beyond science? My answer is yes.

I include these matters in a chapter on faith and science because Marx represented an attempt to turn science into a faith. His insistence upon action is not a moral injunction in the usual sense, but it behaves like one. It demands that we act correctly in answer to a higher power. In Marx this happens not to be God, but his idea of science. Whatever we may believe we are subject to this higher power and the only way to behave is in accordance with its laws. Not to behave thus means simply that one will be crushed by its logic. It was that last conviction that was to justify the worst of all the horrors of the twentieth century – the Stalinist Terror in the Soviet Union.

It is sometimes assumed that the discrediting of the various experiments with Marxism has discredited both the Marxist idea and the idea of a scientific society. This is not so. In the first place the idea that we can evolve a science of society and politics is still alive in much thought of both the right and left wings of politics. Secondly, the fall of Marx did not destroy the idea of a scientific society, it actually made it possible. As I said in my first chapter, liberalism is the true scientific society and it is liberalism that has economically defeated communism.

The other path from Hegel – and the one that attempted to preserve the outlines of the Christian faith – lay in the development of liberal theology. This amounted to a refinement of Protestant individualism combined with the Hegelian theme of the narrative of history. The great story of the 'world spirit' could be examined by the individual conscience. The revelations of science were part of this narrative and faith had to be alert to the process to grasp the divinely inspired plan. In our own time the sheer attractiveness of the idea of liberal theology accounts for the huge success in the sixties of the Catholic thinker Teilhard de Chardin. He devised an elaborate system for explaining the development of human knowledge in the context of our movement into the 'noosphere' in which the act of knowing would predominate. The easy attraction of such thinking is that it effectively says there is no problem, everything is included, all will come right in the end.

Liberalism in theology also lies behind the pervasive cultural liberalism of our day. The revelation of the world spirit was a universal revelation, available to all people and all cultures rather than specifically available to Christians. All cultures could thus be said to be equal in the eyes of this universal, evolutionary history. Again there was no problem.

But there is a problem. For liberal theology it lies in the difficulty of retaining any meaning at all in its religious foundation and in convincing anybody that it is anything but wishful thinking. Liberalism happily accepts any number of increasingly non-literal interpretations of the Bible while trying to preserve the reality of the underlying theology. Indeed, liberal theology is actually defined by its attempts to understand how exactly any such reality can be established. It is, in this sense, a project designed simply to have the best of both worlds.

Two dangers arise from such an approach. The first is the extreme subjectivism it can inspire. This was, of course, implicit in the Protestant emphasis on the individual conscience. But liberalism accelerated this inward movement of the faith by encouraging an understanding of religion as inwardly rather than outwardly determined. In Paul Tillich, the modern Protestant theologian, this process finds its logical expression when he advises people to speak of God as 'of the depths of your life, of the source of your being, of your ultimate concern, of what you take seriously without any reservation.'[13]

It is difficult to imagine a more feeble attempt to save the deity, nor one that more clearly demonstrates the damage done to religion by science. All Tillich finds that he can say of God is that he is a condition of moral seriousness. This is Kant made simple and undemanding for simple and undemanding minds. God requires nothing of us except that we think deeply about things.

The second danger of liberalism lies in its permanent vulnerability to its own analytical thought. Liberalism requires theological activity, constant examinations of the faith. But questioning the details of the faith always carries with it the danger of questioning its entirety. This is precisely what happened to German theologians in the nineteenth century. It was a critical phase in the history of European belief.

Inspired by the work of Kant, Lessing and, most importantly, Schleiermacher, these first truly modern theologians applied critical thinking to what really constituted Christianity. Schleiermacher began the process of centring the faith on man. For him, man's feelings were the grounds of reality and Jesus represented simply the man whose feelings had attained the highest degree of perfection.

In the nineteenth century such critical thinking was applied to the Bible and, in the process, its fundamental authority was effectively destroyed. The most celebrated assault came from David Friedrich Strauss who produced in 1836 *The Life of Jesus, critically examined*. Strauss's problems with Christian orthodoxy arose directly from a study of Hegel. The idea of an evolving world spirit left him puzzled as to what orthodoxy he could honestly teach. His book, which dealt with the Christian drama entirely in mythological terms, was the seminal expression of the non-literal interpretation of Christianity. It may be said to be a central text of the nineteenth century. On translating it into English the novelist George Eliot lost her faith. It destroyed Strauss's career as a teacher, even though his entire effort had been directed towards the recovery of Jesus as a meaningful figure.

From this sprang the radically tragic Christian theology of

Albert Schweitzer, who regarded Jesus as being positively wrong. The Kingdom of God had not been at hand. He died forsaken, having believed his own sufferings would at once transform the world. The only Christ left at the end of this process was an exemplary figure of overwhelming moral stature and nobility. But God? No.

Such a process can too easily be seen as a long retreat. In the case of Tillich, it clearly was. First we found that God was not 'up there' and Hell was not 'down there' in any literal sense, then he had been evacuated from our world entirely. Next we looked inward to discover him there. But, simultaneously, the other outward, historical manifestations of his existence were being wrecked by critical – i.e. scientific – thinking. Finally, the Christian is left with nothing but a good man and an inner moral seriousness.

No wonder Nietzsche was roused to fury. There was nothing left worth defending. It was time to start again in suffering, rage and heroism.

Or perhaps not. For me the Protestant effort did produce one overwhelmingly *convincing* defence of the faith *as a faith*. I stress the word 'convincing' because I do not necessarily mean persuasive, I mean coherent and unarguable on its own terms. A defence of religious belief does not necessarily have to convince others, it might merely mean the discovery of a way in which it can be coherently sustained in the defender. And I stress the phrase 'as a faith' because this great act of defence, or rather defiance, turned upon the deepest meaning of the word 'faith'.

For what does 'faith' mean? Clearly it cannot mean being rationally persuaded of something. If we had a reason for faith, then it would not be faith at all, it would be logic. Faith can only be unreasonable.

Furthermore, Christianity, when stripped of its medieval accretions, actually insisted upon this unreasonableness. It had always been a religion of redemption through suffering. Its nature was paradoxical, irrational: to find the light you must pass through darkness, to find peace you must endure turmoil, to attain everything you must first have nothing.

In this context it might be possible to turn the cruel invasion of science into something else. It might be just one more trial through which the faith had to pass. The destruction of 'evidence' for the faith by the extension of our scientific knowledge might simply be a way of driving back on faith itself. Perhaps the very impossibility of belief was the point.

This was the solution of Sören Kierkegaard, the greatest theologian of the modern world and possibly the one man with an intensity of mind to match the destructive atheism of Nieztsche. Upon the extremity of the modern demand for unbelief, he constructed his faith. In his short life (1813–1855) he defined the ultimate position of Protestant individualism. He rejected Hegel's metaphysics as too easy, a philosophy that removed responsibility from the individual for his own life and his own choices. He rejected also the subtle abstractions of the idealistic Christian apologists.

To Kierkegaard the world demanded of the individual one highly specific act: a choice. This 'authentic choice' applied to all areas of human life, but, most importantly, it applied to faith. Christianity was not a reasonable proposition, it was not likely to be true, one could not arrive at such a system by rational processes. One must *become* a Christian. Christianity cannot be plausible; if it were it would be the softest of options, a mere choosing of the nicest alternative. But Christ suffered on the cross and the choosing Christian would have to endure comparable sufferings. Above all, he would have to suffer the sheer improbability of what he had to believe. One becomes a Christian in spite of everything. The effort of that becoming, the struggle against the claims of rationality, lay at the authentic heart of the faith.

The importance of Kierkegaard was that he found a way of turning the retreat of religion into an advance. He required neither subtlety nor evasion to sustain his faith, he simply required the reality of choice. This stripped away the details of theological and philosophical debate as well as the material inroads of science into the spiritual empire of the Church. It even eliminated the entire issue of the meaning of God's 'existence'. We chose and that was that.

But, if this were a triumph of the intellect and of courage, even if it were 'true', it could scarcely offer a programme for the revitalization of the Church as an institution. Kierkegaard was too demanding, too individualistic and too removed from the literal realism that people demanded of their faith for his teaching to become popular. In addition, of course, the fact of the choice meant that you could choose against faith – only by retaining that possibility could you retain the authenticity of the choice.

As a result of all these factors his legacy, perversely, has been the most recognizable image of radical atheism in our own time. For Kierkegaard created existentialism, the popular modern philosophy of the individual as isolated with his own choices, creating himself anew each day. Pessimistic and narcissistic, the existentialist becomes the hero of his own story, the one self-created object in his world. It was not a legacy he would have liked because it represents precisely the wrong choice.

Kierkegaard's central theological argument springs from the imagination of an artist. The literariness of his works is as important as their theological content. They are as much expressions of the problem as arguments about it. In the end he was asking us all to have souls as great as his own. No demand can more precisely oppose a man of genius to the modern world. He was certainly at odds with his own age. Indeed, he consciously fought against his time, even deciding not to marry because to do so would be to tie him too precisely to the history of the nineteenth century. But, in spite of his sheer oddity, I believe Kierkegaard's importance lies in the clarity with which he saw the issue. He saw that our humanity could only be saved by an act of absolute assertion, of choice. This choice was made on the basis of our selves in spite of, even in opposition to, the facts of the world. Perhaps his demands seem extreme only because they came too early – perhaps now we can see they are not extreme at all, merely necessary.

But the history of his own time was, of course and in spite of everything, the history of the relentless progress of science. In the first half of the century the residual theology of the Enlightenment remained. The argument from design – God as the supreme engineer of all this newly discovered order – still held the imagination. But, as the industrial revolution progressed and science became a powerful, professional institution, God retreated popularly as he had already done intellectually.

The nineteenth century, however, was not simply, or even primarily, an age of heroic technology fabricating a new world and intellectual giants struggling to discover a new world order or salvage the remnants of the old faith. It was also the age of a new type of man with a new type of faith, a faith that embraced science as a myth of progress and improvement. In terms of the increasing numbers of literate people, this was the authentic new faith that really did replace religion with its demands, its darkness and mystifications. For the nineteenth century was also the age of Homais the chemist.

Homais is perhaps the greatest character in, for me, the greatest of all novels – Gustave Flaubert's *Madame Bovary*. Published in 1857, the book is a conventional tale of provincial adultery. A passionate, selfish, imaginative wife turns against the narrowness of her existence, married to a dull husband, a doctor, and confined by the mores of a small town. She is destroyed by the process and dies hideously from self-administered arsenic.

Homais is the local pharmacist, the possessor of the arsenic. He is a New Man, a prophet of the age to come. He is a liberal-minded sceptic, anti-clerical and progressive. His being is a celebration of the peak of nineteenth-century civilization, rather like the Natural History Museum in London. History is an open book to Homais, he knows everything. Science, he is convinced, will one day solve all our problems. In all this he is rather like us. For Homais is indeed *the* New Man, an embodiment of all the beliefs of the new age. The catch is that Homais the chemist is the most vivid, unforgettable realization of evil ever to spring from the Western artistic imagination.

Flaubert's genius in creating this colossal monster was to show, from one perspective, the terrifying inadequacy of his beliefs and yet, from another, their equally terrifying adequacy. Nothing in Homais's glib doctrines can ever provide meaning or consolation for the passions of Madame Bovary. Confronted with human weakness and imagination, Homais can only respond with the bland, supercilious assurance of the technocrat. From the perspective of suffering humanity he has nothing to say. And yet, from his own perspective, he can say everything. He *knows* that all this drama is just a passing, local fragment of history. He *knows* it has no meaning other than within the impersonal narrative of progress. Homais, after all, is a scientist, a technologist, a sober, serious member of the community. Homais can keep things in perspective. Homais is a bourgeois.

The appalling tragedy is that Homais is right. Madame Bovary is an inadequate. She lives in dreams. Certainly she is an artist – *'Madame Bovary, c'es moi,'* Flaubert said – but, in this new world, to be an artist is to be a sideshow, a passionately yearning creature with nothing to yearn passionately about. She is meaningless, whereas Homais is replete with meaning.

Society understands and rewards this. The last line of the book tells us that Homais had been awarded the Legion of Honour while Emma lies mouldering in her grave. It is a line that is designed to evoke in the reader an unbearable sense of rage and injustice. But it is also a line designed to make us wonder what we are angry about. Homais is a vile, social-climbing, inhuman monster. But what have we to offer that is so much better? Our rage is as baseless as Emma Bovary's yearning.

Flaubert turned the rage into his art. As Irving Babbitt has written Homais was his vision of 'contemporary life and the immeasurable abyss of platitude in which it is losing itself through its lack of imagination and ideal. Yet this same platitude exercises on him a horrid fascination. For his execration of the philistine is the nearest approach in his idealism to a positive content, to an escape from sheer emptiness and unreality.'[14]

Impotent rage and sick fascination provide a kind of affirmation in the face of the world of Homais, the bourgeois.

This digression into fiction is an attempt to describe what was at stake imaginatively in the nineteenth century. Primarily it was the age in which the full personal, social and political implications of a triumphant culture of science were finally realized. The decline of faith and the oppressive sense of fragmentation that accompanied this realization meant that it was an age littered with elegies for a harmonious past of faith and meaning. Romantic art was full of medieval landscapes, primitive, 'organic' cultures and the peace of unspoiled nature. But it was also littered with people like Homais, prophets of the new progress.

In the middle stood men like Kierkegaard and Flaubert, the first trying to make the present work as the present by providing it with a modern theology and the second raging in despair that the modern was not worth having and yet it was all there was. Both were vicious and implacable enemies of the bourgeois faith.

For the bourgeois is the central character in the post-religious, scientific drama. He might be said not to exist as a real individual except in the demonologies of these great souls who saw meaning draining from the world. But he unquestionably exists as an aspect of us all, a fundamental type of the present.

The precise resonance of the word is important. The bourgeois is not merely middle class, nor is he merely an anti-clerical technocrat. He is not merely materialistic, nor is he merely complacent. He is all of these and yet he is also savage and inhuman in defence of his own complacency.

Certainly he is shallow, but his roots run deep. These roots run back to the new merchant class that sprang up in the fifteenth and sixteenth centuries. The success of this class might be said to be, like science, based on an essential amorality. For, like science, trade appealed to an external value that was not religion. In this case the external value was the demands of trade itself, later to become the whole, elaborate structure of economics.

The symbol of this value was usury – the earning of interest. Usury was a subject of profound dispute in the Middle Ages precisely because of its obvious amorality. For usury says that money has inherent value. It does not have to do anything to be worth something. Money simply lying in a bank earns interest.

R. H. Tawney wrote of this enormity: 'To take usury is contrary to scripture; it is contrary to Aristotle; it is contrary to nature, for it is to live without labour; it is to sell time, which belongs to God, for the advantage of wicked men; it is to rob those who use the money lent, and to whom, since they make it profitable, the profits should belong . . .'[15]

Usury was, above all, irrational in the context of a Thomist world. It turned money into an abstraction passing judgement on the activities of the world. Any project based on borrowed money could only be judged by its ability to repay the interest. Against the systematic rationalism of the medieval world, the Renaissance merchants set up this irrationalism based on the arbitrary ascription of an unchallengeable value inherent in notes, coins and, in a further irrational refinement, the solemn assurance that such notes and coins could be exchanged for nothing more solid than a bank statement.

Usury was thus irrational to a medieval mind in precisely the same way that science was irrational. It was abstract, subjective, arbitrary, far removed from the natural facts of the world. Such qualities appalled rationalist intellectuals and their struggle to base economic value on firmer foundations lasted well into the nineteenth century. Karl Marx with his labour theory of value – an attempt to see money as the precise correlative of work – may be said in this context to be the direct descendant of Aquinas. Aquinas wanted a consistent intellectual basis for his faith of Christianity; Marx wanted a consistent intellectual basis for his faith of economics.

But the mercantile imagination, fired, like science, by its own success, cared little for such refinements. Indeed, it cared little for any issues normally categorized as religious. 'Everywhere,' wrote Max Weber, 'scepticism or indifference to religion is and has been the widely diffused attitudes of large-scale traders and financiers.'[16] Trade seemed to give an objective rationale for human existence that reduced the need for belief. Science, to the complacent merchant, appeared to validate this scepticism. So science impregnated trade and, from its smug womb, the bourgeois was born.

When Homais first appears in Flaubert's novel, he is seen from a distance bent over his desk. His home is described, plastered with advertisements for patent remedies – blood purifiers, Regnault's ointment and so on. On the shop front is the sign: HOMAIS, CHEMIST. Inside there is a further sign: LABORATORY. The commerce and the science are one in the house of Homais.

The problem with the bourgeois was that he saw no problem. Indeed, the phrase 'No Problem' is the motto of every contemporary Homais. Emma Bovary's tragedy is meaningless to him. She brought it on herself with her own silliness. Observing this, all the art and genius of Flaubert can do is heighten our disgust, fire our loathing. Similarly Kierkegaard was driven to heaping such immensities of moral responsibility on to the individual soul that no bourgeois could possibly take the strain. Nietzsche merely demanded superhuman courage, vision and aristocratic disdain for the suffering of others. All were appalled by the complacent bourgeois compromise they believed was threatening the spiritual health of the species; indeed, threatening the very existence of the spiritual.

They were, of course, right. The twentieth century may have stripped the bourgeois of some of his progressive ideals, but, in essence, his faith in the combination of economic growth and scientific rationality has become the underlying religion of our age. Other beliefs may be held and other doctrines propagated, but this is the only one that can be said to be a necessary characteristic of our modern civilization. Homais has triumphed.

And this triumph has marginalized the descendants of the prophets who rejected the values of the bourgeoisie. For the rise of the bourgeois created the intellectual. Stripped of religion, the anti-bourgeois had to seek other rationales for his loathing of Homais.

'The intellectual seeks in various ways,' wrote Weber, 'the casuistry which extends into infinity, to endow his life with pervasive meaning, and thus to find unity with himself, with his fellow men, and with the cosmos. It is the intellectual who transforms the concept of the world into the problem of meaning. As intellectualism suppresses belief in magic, the world's processes become disenchanted, lose their magical significance, and henceforth simply 'are' and 'happen' but no longer signify anything. As a consequence, there is a growing demand that the world and the total pattern of life be subject to an order that is significant and meaningful.

'The conflict of this requirement of meaningfulness with the empirical realities of the world and its institutions, and with the possibilities of conducting one's life in the empirical world, are responsible for the intellectual's characteristic flights from the world.'[17]

The intellectual is one who cannot collude with the blank simplicity of the bourgeois world view, with its easy progress and its all-conquering science. So he seeks his systems to show that the world is more elaborate, finer and more inclusive than anything in the dreams of Homais. But the effort seems futile, first because all his systems are inventions, fictions, works of art. They have nothing to compare with the simple bourgeois certainties. And secondly, even if they did attain comparable

certainty, they would remain in the marginal realm of the intellectual – in the smart, café society that has characterized the modern
intellectual life. Every literary clique, every artistic set, every
tasteful fad is a continuing expression of the sterility of the role
the intellectual has taken upon himself.

For the truth is that what the intellectual quest really needs is
a religion and yet it is fundamental to the nature of intellectualism
that that is the one thing the intellectual cannot have. He can
neither embrace the old faiths, nor can he invent new ones. All
his ideas are condemned to pass their time on the margins of a
culture that has chosen its own faith, its own metaphysic and
which has no need of his refinements.

So, by the end of the nineteenth century, the prevailing
religious orthodoxy was clear. In its bourgeois form it was the
pragmatic unity of science and trade. Beyond this lay the moral
cosmology that science seemed finally to have completed: that of
the meaningless universe. Man, in Freud's summary, was alienated from the universe, nature and himself. Religion no longer
accompanied the highest and best of human thought. Instead it
had become one more object of scientific curiosity. Either it was
an obvious mistake, an intellectual error. Or it was a symptom
of human discomfort and discontent – illusory fulfilments, in
Freud's words, 'of the oldest, strongest and most urgent wishes
of mankind'[18] Against that there was the contrast of the form of
knowledge offered by science – 'the only road which can lead us
to a knowledge of a reality outside ourselves.'

Freud recognized the bleakness of such a conclusion as well
as his own role as a modern incarnation of the sorcerer, but one
without magic: 'Thus I have not the courage to rise up before
my fellow men as a prophet, and I bow to their reproach that I
can offer them no consolation: for at bottom that is what they
are all demanding . . .'[19]

In the place of religious passion there could now only be a
kind of hopeless urbanity. What, the British geneticist J. B. S.
Haldane was asked, could he deduce about the nature of the
Creator from his creation. 'An inordinate fondness for beetles,'[20]
he replied with all the dismissive urbanity required of us by
modern 'sophistication'. There was nothing there, but what was
there – beetles.

Worst of all, there was nothing in such science to replace the beauty and poignancy of the Christian myth. With a clinician's sigh of regret, Freud explained that our need for a single, all-powerful god was nothing more than the human psyche's need for a father.

Religion had been defeated. Western society would, henceforth, be secular. The sheer energy, power and effectiveness of science had weakened the old faith until it had become just one more voice among many others, merely an opinion. It may have answered questions that science did not, but the source of its answers was no longer believed, so neither were its answers. We would just have to live without those kinds of answer – or pretend to provide them from the safety of our new posturing, smug roles as intellectuals or bourgeois.

Of course, we still preserve the language of the old faith at Christmas or in the desperate demands of the American television evangelists. Most commonly we choke with nostalgia at the thought of the certainties it must have provided.

But, even in the midst of our most fervent nostalgia, we knew that the past was never as easy as science and technology have made the present. The effectiveness of science weaves its familiar, seductive spell. Whatever this appalling, comfortless knowledge meant, we could not deny it worked. It made bourgeois of us all. The problem was that it left us with the aching, anguished loneliness of scientific man in a universe which, in some ghastly parody of the original fall from grace, his knowledge had stripped of goodness or meaning.

The defence of the faith had failed and the soul of modern man had been formed. In 1869, after Newton, after Darwin, after Strauss, after Kierkegaard, after Flaubert and with Freud already growing up in Freiberg, the English poet Matthew Arnold looked out upon the sea at Dover. The sound of the waves on the shingle was a 'melancholy, long, withdrawing roar' which seemed to him like the sound of the sea of faith retreating from the earth. It left behind only the mutual consolation of human beings in the face of the meaningless world of the ichneumon wasp. The beauty and wonder of creation that had inspired the argument from design in reality told us nothing. Beauty was not truth, truth was not beauty. All that was left was the private avowal in the face of the unspeakable.

> Ah, love, let us be true
> To one another! for the world which seems
> To lie before us like a land of dreams,
> So various, so beautiful, so new,
> Hath really neither joy, nor love, nor light,
> Nor certitude, nor peace, nor help for pain;
> And we are here as on a darkling plain
> Swept with confused alarms of struggle and flight,
> Where ignorant armies clash by night.

5

From scientific horror to the green solution

For each of our actions there are only consequences.

Lovelock[1]

Today we are faithless, but pampered by the affluence that scientific society provides. As never before the means to live come easily to most of the human species.

A supermarket is an elegantly simple and clear image of this affluence and this ease. Its logic is exquisite. We have cars, so why walk to a number of different shops when we can park outside just one? And we have so much to consume now that it no longer makes sense to ask for each item by name from a shop assistant. Instead we are permitted to wander freely among the loaded racks, shelves and gondolas, filling our wheeled baskets.

When supermarkets first appeared in Britain in the fifties they felt like a warm breeze from the future, from the infinitely privileged, sadly distant, western shore of the Atlantic. They embodied the dream of abundance and easy availability and the certainty that at least one type of human problem had been solved. Feeding ourselves was no problem.

But in the last few years supermarkets seem to have started to advertise more than just abundance, availability and cheapness. Now another justification has appeared for why we buy, how we buy or what we buy.

Here is a supermarket today. There are two different types of vegetables – the usual and those that are 'organically grown without chemical spray or artificial fertilizers'. Here are garbage bags made with '50 per cent minimum recycled plastic'. Here is toilet paper made from 100 per cent recycled paper – 'Helps save trees,' explains the packaging. Here are 'skin, hair and bath preparations specially formulated with plant, fruit and flower extracts, produced without cruelty to animals'. Here is a soap-powder box with, printed on the side, a panel of 'Environmental Information' about biodegradability, energy usage, packaging and 'gentle' stain-removing chemicals. 'Support recycling in your community,' it says at the end of the panel. Here is Greencare chlorine-free bleach – 'Towards a cleaner environment'. Here is hair-spray which is said to be 'ozone-friendly'. And here is a leaflet from the government's Department of the Environment, cheerily calling on us to 'wake up to what you can do for the environment'. Inside are solemn injunctions on how we must change our lives by using low-energy light bulbs, disposing of old fridges carefully, reusing plastic carrier bags, planting trees, using unleaded petrol, not uprooting wild plants, not dropping litter, scooping up after our dogs and checking on river pollution. The leaflet is published on recycled paper.

Supermarkets live or die by selling us things. Every square foot of floor space, every can, every packet, every word on every wrapper is focused on that moment when the red laser beam scans the bar code and what was theirs becomes ours and what was ours theirs. The whole colourful paraphernalia of marketing is aimed at and given meaning by this instant.

In a simple world with no fear, imagination or aspiration, selling would be a rational, finite process. We would buy this because it was cheap, this because it tasted better and so on. In the real world, however, selling is obliged to match the complexity of the buyer's psyche. The salesman must be infinitely patient, pandering to the customer's myriad irrationalities. To sell somebody something in a competitive, rich society you must, in an important sense, know the buyer. And what all these strange messages about recycling, CFCs, low-energy and pollution signal is that the supermarket knows its customers well enough to know they have discovered some new justification. They have found a rationale, a grand simplification that will lead many of them to make buying decisions they would never have made before.

The justification is the protection of the environment, the 'Green' impulse that has captured the imagination of the developed world and modified the easy objectivity of the market place. The importance of this impulse is that it attempts to inject an entirely new set of values into the scientific society from outside the logic of science and liberalism. The need for this injection springs directly from something quite new that happened to science in our century – it ceased to be trusted by the very people on whom it was showering the greatest material benefits. Its effectiveness was finally questioned.

'We are the hollow men,' wrote T. S. Eliot, '. . . . our dried voices, when/ We whisper together/ Are quiet and meaningless. . . .'[2]

The dark shadow of the pessimism that fell over the climax of the classical scientific project has proved even darker in our century. In the face of the vision of man as a fragile, cornered animal in a valueless mechanism, the modern imagination in art and literature has often decided it could do no more than build artificial castles in the rubble of the culture and a general cultural pessimism, even despair, has been a commonplace of educated society. In modernism artistic form took precedence over content for there seemed, in some terrible way, to be no content left. In observing himself and his plight – the traditional concerns of art – modern man could only see a great vacancy.

Eliot's poem *The Hollow Men* (1925) has an epigraph – 'Mistah Kurtz – he dead.' – from Joseph Conrad's novella *Heart of Darkness* (1902). Eliot was bowing to Conrad as the creator of a modern myth of rare intensity. It was a book that was clairvoyantly to capture the pain of the twentieth-century legacy in a single narrative image; its brevity and succinct power make it an unsurpassed evocation of a world in which meaning cannot be found.

Heart of Darkness concerns the quest for Kurtz, a trader in the African jungle. When found, deep in the darkness, he has undergone a revelation which amounts to no more than, in his dying words, 'the horror, the horror'. The hero returns home to tell Kurtz's beloved of his end. Confronted with her grief, he finds he must lie. He tells her that the last words Kurtz uttered were her name.

For the secret catastrophe of the modern mind is too terrible to be acknowledged in polite society. Human beings cannot live with such a revelation. The only morality left is that of the consoling lie. In the absence of great old illusions, little new ones must be our consolation.

High art, however, has its preoccupations; they need not necessarily reflect those of the majority. Yet it is one of the many strange ironies of our age that, just as the material benefits of science and technology were invading our lives in force, transforming our economies and our health, the terrible vacuum beneath their achievements was suddenly being made dramatically apparent to us all. The horror glimpsed by Kurtz was to be evoked in detail on the nightly news. What had once been a rarefied unease becomes, in our age, a popular dismay disseminated and nurtured by patterns of electron vibrations.

Inspired by its mythic directness, Francis Ford Coppola used Conrad's novella as the basis for his Vietnam War film *Apocalypse Now*. As the century passed, the myth of inconsolable despair and visionary horror seemed to have become ever more appropriate. In Coppola's version our entire civilization was tried and convicted in the jungles of South-East Asia. A futile struggle for 'democratic values' had become a blood-soaked festival of random death. And we knew that it was true because we had seen it on TV. Coppola simply extended and intensified the images we already knew, the despair and nihilism we already felt.

Our century has provided a vast spectacle illustrative of both Conrad's and Coppola's horror as well as of Eliot's cultivated anguish. Only the most wilfully insensitive could be unaware that something has gone badly wrong with the nineteenth century's dream of material progress. For we have not only inherited that century's legacy of the cold shock of a meaningless universe, we have also to cope with the discovery of a range of potential evils unknown to the world before the advent of science and technology.

It should not have come as a surprise, of course. As I have shown, the severance of knowledge and value had been a commonplace of philosophy for two centuries. But, somehow, the terror of the idea seemed to require an image to project it out of the studies and salons and into the world, an image that would demonstrate the awful truth that what we knew had not redeemed us from what we were. It needed to be more than just a theory, like, for example, Darwin's powerful evocation of the kinship of men and apes. It was shocking, horrific even, but the popular mind could still smile at and ignore the idea. What was needed to make the point was a concrete realization. Our age, being scientific, requires 'hard evidence'.

We were, in the event, to be deluged with such evidence. The first, and still the most potent example to those who experienced its devastation, was the First World War. The Great War revealed the black heart of industrialization just as surely as the concentration camps did in the Second. And the climax of the Second, the atom bombs dropped on Hiroshima and Nagasaki, suddenly revealed science itself as an uncontrollable extension of the human will to destruction.

A mechanistic science was clearly not enough. This had been understood by the philosophers of romanticism and the Enlightenment. But the twentieth century saw the point driven home with relentless brutality. We all now began to realize that science could not, in Max Weber's words, respond to the 'inner compulsion to understand the world as a meaningful cosmos and to take up a position towards it'.[3] Furthermore, the atom bombs demonstrated that science may not merely be deficient in this respect, it might be positively hostile. It was one thing to say with Freud that no answers in the realm of meaning and value could be provided, quite another to show that, if no answers were forthcoming, then science would become a tool of overwhelming power placed in the hands of a species that remained spiritually infantile.

The enormity of this shock – or series of shocks – is intensified by the spirit of immense, material optimism with which our century began. In the last decades of the nineteenth century and in the first fourteen years of the twentieth, technology was working furiously and with unprecedented productivity to create practical outlets for the advances of science. At last the legacy of Newton's and Galileo's cosmic understanding and of Descartes's diligent method began directly to change the lives of whole populations, apparently for the better. People were being involved in science at the level of their daily lives. It was as if they were being deliberately prepared for the coming cataclysmic disillusion.

There was an acceleration in the societies that possessed

science; space and time were found to be compressible by human ingenuity. Communications had depended on line of sight or the speed of horse or foot for the whole of history until at first steam, then the car, aircraft, the telegraph, the telephone, radio and television transformed our sense of distance and of our own limitations.

The progressive ideas of the seventeenth and eighteenth centuries had, without adequate technology, remained in the realm of the imagination. Progressive notions of growth were restrained by the way in which science, successful as it was, remained unable to move from the explanatory to the practical. But modern technology suddenly made the dreams realizable. The ideal of economic progress as the goal of all our labour spread to a new and rapidly more affluent population. The curve of productivity growth steepened as primary mechanization gave way to the systems of mass production and as sophisticated financial mechanisms emerged to provide investment capital.

The expansive mood of the time encouraged a search for global solutions. Most spectacularly and influentially, on the basis of Karl Marx's pseudo-scientific metaphysics and Lenin's metallic conviction, the Tsarist Russian Empire became, in 1917, the communist Soviet Union. A momentous transformation in the life of a vast nation had been executed on the basis of a single theory that was said to be so utterly scientific that it justified any human sacrifice. To Lenin human lives presented a 'problem' to which quasi-scientific analysis, technology and correct political action could provide a 'solution'. In time, the logic of the dialectic assured the revolutionaries, the rest of the world would follow where they led. This was what they thought science had told them.

Technology, meanwhile, had attained critical mass and begun to explode in all directions. The body of science available to the world by 1900 was enough to power a bewildering number of practical applications. In 1895 the novelist Henry James had electric lighting installed, in 1896 he rode a bicycle, in 1897 he wrote on a typewriter and in 1898 he saw a cinematograph. Gadgets were everywhere and, by the 1890s, domestic electricity became widely available to make them work.

There was a material confidence in the air that contradicted and yet was the necessary correlative of the mechanistic despair of the nineteenth century. The despair was clearly a luxury, an insight available to those who had the time and the means to savour its implications – available, certainly, to Conrad and Eliot. But, if practical science could solve the problems of the poor and oppressed, then, to the practical man, immediate benefits would outweigh any spiritual crises. There was too much to do to worry about the condition of one's soul. Despair was the prerogative of the intellectual; technological optimism that of the bourgeois.

So, across the Western world, the forty years up to 1914 were a period of extraordinary growth and prosperity. Europe and America were rapidly becoming urban continents with a sophisticated urban conception of what was possible. Economics prospered as the science of the management of growth. It was supported by statistics that had become more reliable because of a more ordered, industrial landscape. The nation state that had been born of the secularization of the Enlightenment was now being corporatized and its performance monitored. People began to believe, on the basis of this vast new source of information, in a 'scientific' sociology and politics. Above all, it was an era in which the issues of the modern began to press in on the human imagination. Indeed, it was an era in which the very word 'modern' was given meaning by the pervasive expansion of possibilities and the certainty of change.

'It is probably fair to assert', the historian Norman Stone has written, 'that Europe, before 1914, produced virtually all of the ideas in which the twentieth century has traded; the rest being mainly technical extensions of these ideas.'[4]

This is not quite true. Quantum mechanics, for example, came later. But there can be no doubt that it *was* an age when science and technology appeared to be coming to a benevolent maturity and the belief of men like Lord Kelvin that physics might indeed be nearing completion as a subject reinforced this sense that human knowledge was attaining some form of apotheosis. At Harvard University in the 1880s John Trowbridge, head of the physics department, was telling his students that it was not worthwhile to major in physics as every important discovery in the subject had now been made. All that remained was a routine tidying up of loose ends, hardly a heroic task worthy of Harvard graduates.

The confidence was justified in the limited sense that technology had indeed reached an evolutionary take-off point. We now take an all but infinite diversity of innovation for granted. The practical realizations of scientific theory are routinely understood to be, in effect, limitless. This sense of the infinite potential of human ingenuity was born in these climactic decades of the long European peace of the nineteenth century. It was to lay the foundation of our second industrial revolution and to make the twentieth century the First Machine Age.

In almost every other respect, however, the confidence was fatally misguided. It was misguided because it was soon to transpire that physics was far from complete; indeed, as a subject, it had scarcely begun. But it was also morally misguided. For, as the First World War was to reveal, the acceleration was not exclusively confined to the humane benefits of technology. Napoleon's armies had moved no faster than Caesar's, but those of the Kaiser were mechanized. Airships and aircraft were able to carry destruction far beyond the battlefield and the immediate area of conflict. Wars could now become as global and lethally efficient as the communications that reported them. In one day of the Battle of the Somme in 1915 more lives were lost than in the whole previous century of conflicts in Europe.

The economic maturity of the nation states had evidently not been accompanied by any moral transformation. In the trenches of the First World War the age's buoyant image of itself flipped over to reveal an underside of cultural despair. A new age of irony was born in the conflict between the image and the reality – at Ypres, the Somme and the Marne. It was a modern form of the irony of Galileo, finally old and blind, or of Newton, the Grand Wizard, as the boy on the beach. It was the irony that has always accompanied progressive, Western man. It springs from the simple contradiction between the power of our imagination and the vulnerability, inadequacy and mortality of our bodies.

In the trenches such ironies were plentiful. The past glories of military conflict, which had fired the minds of the young men who left for Flanders, were transformed into a mud-bound stasis.

The age of machinery and mobility had produced a war of total paralysis. The territorial demands of defence and attack were reduced to an inept, catastrophic squabble over a landscape made worthless by high explosive. The dream of progress was sublimated into a nightmare of immobility. It was a bad joke.

Modern death was also born in the trenches. Modern death is a meaningless, organic failure. The shells land where they may and the bodies are torn apart. They are machines, so, when broken, they stop working.

It was all a hard, bitter lesson for the liberal ideas of progress that had been nurtured by the success of the Enlightenment. All this science, technology and wisdom, the apparently limitless powers of human reason had come to this. There had never been a war like it. Two centuries of hubristic humanism were entombed in mud.

'The world,' wrote Irving Babbitt in 1919 in a dry comment on the foundering of the Nietzschean dream, 'it is hard to avoid concluding, would have been a better place if more persons had made sure they were human before setting out to be super-human.'[5]

'Was it for this,' asked the English soldier-poet Wilfred Owen as he surveyed the devastation, 'the clay grew tall?'[6] And Ezra Pound, an American who had flung himself physically and intellectually into European culture as if for his own salvation, saw the young men 'believing in old men's lies' and dying 'For an old bitch gone in the teeth. For a botched civilization.'[7]

It was the Second World War, however, which was to make clear, first to the warlords and later to the population as a whole, that science and technology as destructive forces had come of age. It was the first war in which air power and rapid battlefield movement meant the inevitable and total involvement of civilian populations. Bombers struck at cities, not because they contained combatants, but because they contained industries supporting the war machine as well as people whose morale could be destroyed. The rationalism of modern war demanded that victory or defeat were no longer technical matters to be decided by opposing armies, but final conclusions about the viability of a people and its way of life. The climax of the European nation state was this technologically inspired discovery that to have a nationality was to be part of a military process. Women and children endured passive recruitment.

This totality of war applied also to the kind of solutions it was supposed to be able to provide. On the one hand civilian populations could be bombed in order to advance military aims. Yet, on the other, they could be assaulted in the service of a higher rationalism and morality that could cancel mere pity. Nazism is meaningless without its central conviction that it was *good* for people.

For it was the war of 1939–45 that finally established the idea of war as the tool of a higher, grander plan. The logic of the German strategy was determined by industrial capability, but it was spiritually underpinned by a Nazi pastoral ideal of health, fitness and moral cleanliness. The perfect Aryan male was, by some strange technological alchemy, to be transformed into the flawless mechanized warrior with his vast and lethal prosthetics of tanks, guns and, most glorious of all, aircraft. Flight brought with it a sense of Nietzschean purity, an inhalation of cold clear air in the heights from which the behaviour of lesser men could be surveyed with scorn.

Combined with Hitler's appeal to German nationalism, this produced a murderous totality. This state, this race was different, superior. It was not a case of Christians fighting Christians or quarrelling monarchs settling their differences on the battlefield. It was a case of a new perfection cleansing the old impurities. The Final Solution – the Holocaust – was logical enough in such a context, a purging of the homeland's racial stock. Yet the shock to the European imagination derived not simply from the fact that people died for such a cause, but from the way they died. The killing was industrial and performed with cold rationality. The Jews were a disease. They would be eliminated with railway trains and nerve poisons.

The reasoning made perfect sense except that it was evil, a technicality only detectable if you did not happen to be a Nazi. The murderers of the SS could obviously have been recruited from the psychopaths in society. But, equally, there would have been the sane, reasonable types who would have concluded that drastic situations required drastic remedies, and perhaps this was that situation and these were those remedies. Rationality is defined by the imagination of the rationalist.

And Germany was the nation of Leibniz, Kant, Beethoven and Einstein. The cradle of the Enlightenment had become the Kingdom of Hell.

'The country in the world', wrote Hugh Thomas, 'with the best education for the longest, the nation with the most serious national preoccupation with learning, the people with the highest rate of literacy in the world in the eighteenth century were the authors of Auschwitz.'[8]

Rationality, technology and the whole Enlightenment project had been shamed. Brute industrialization had turned on its creators. The progressive virtue of the factories and mines of the nineteenth century had once been contradicted only by a humanist horror at the conditions they bred. But it would all be well, progressive people would argue, things were improving. Factories and homes would be better, technology would clean up our lives.

Auschwitz, however, was a factory and Guernica a rational application of technology.

To provide one final twist of the moral knife, there was the spectacle of Nazi 'science'. People and their bodies, living and dead, were used for experiments. There were live medical experiments in the concentration camps and, outside, psychiatrists and doctors selected subnormals, schizophrenics, manic-depressives and alcoholics first to be sterilized and later killed. Neurologists then used the bodies for research and their findings were solemnly published in scientific journals. There is a sense in which this latter process is more chilling than the medical abuse in the camps. For the process separated the scientists from the killing and allowed them to present their results without reference to how they came by the objects of their study. They could plead ignorance or, failing that, the objectivity and neutrality of scientific research.

'Some of the scientists and medical men who carried out this work are still alive,' writes Max Perutz, 'in comfortable retirement, and apparently they look back smugly on their complicity in mass murder . . .'[9]

How could science retain its innocence against such visions?

Perhaps in the bizarre, speculative outer limit of the new physics in which we seemed to be unravelling the inner nature of all matter, scientists could still cling to the essential purity of their task. Nazi killing was tainted with the ancient, choking grime of the Industrial Revolution. The familiar pictures of the extermination ovens evoke disused factories. They could perhaps illustrate a newspaper feature about unemployment in the inner cities. But surely at the heart of the atom lay something pure, clear and untouchable, something beyond those old, filthy mechanisms. On 6 August 1945, even that last hope was smashed.

Hiroshima was, of course, a product of technology and was made possible by industrialization. But these were different kinds of technology and industrialization. Ordinary bombs and concentration camps were obvious, comprehensible developments of nineteenth-century mechanisms. There was no fundamental conceptual shift involved. Similarly the effects of such things could be understood as limited in a fairly ordinary way. Explosives were of a given power, therefore the only way to make a bigger bomb was to make a heavier one and any such development would be restrained by the carrying capacity of aircraft.

But the atom bomb was born of a twentieth-century science. It was only made possible by atomic physics, the most radical and specialized science of the day. It did not explode because of some blending of chemicals but because of a humanly engineered change in the nature of matter. I shall discuss in the next chapter the deeper implications of the new physics. But, for the moment, it is simply necessary to understand that the atom bomb came to symbolize an utterly fundamental transformation in the relationship of mankind to scientific knowledge.

Prior to Hiroshima, science had certainly contributed to the effectiveness of warfare. Mechanization, ballistics, chemistry and aeronautics had transformed and extended the battlefield. But these all remained within the realm of the mechanically comprehensible – in effect, Newtonian killing systems. The bombs that fell on London during the Blitz were understood as the chemical descendants of all other explosive devices and the planes that delivered them were extensions of the guns of the past.

But the atom bomb was incomprehensible in such terms. A single, not especially large, bomb destroyed a city in a fraction of a second. The economics of killing could never be the same again. Crudely one could simply multiply that one bomb by the number that had rained down on Dresden or Hamburg and clearly all human life – certainly all *urban* human life – could be extinguished just as rapidly.

Furthermore, the weapon could not easily be understood as a mechanism. It may have been dimly related in people's minds to the figure of Einstein or to celebrated experimental dramas like Rutherford's splitting of the atom or to Niels Bohr's planetary model of the atom as a nucleus surrounded by a shell of orbiting electrons. But how to make such strange, exalted things into the brutal fact of an explosion was beyond understanding.

Insofar as it could be grasped it seemed to represent some colossal, demonic, human depravity. We had turned the deepest workings of nature into a yet further expression of our inner violence. Meanwhile, the almost simultaneous exposure of what had gone on in the concentration camps left few in any doubt that there would always be men capable of using such weapons for aims less forgivable than ending the Pacific War. The knowledge, the scientific knowledge, was in the world, and it could never again be expelled.

The innocence of the easy, progressive Enlightenment myth, struck down in the trenches, finally died at Hiroshima and Auschwitz. Scientific reason was as capable of producing monsters as unreason. It is no good arguing that Auschwitz was not reasonable – in its way, it was, and secular society cannot be sure that it can offer any higher, purer rationality – nor is there any point in claiming that Hiroshima was a necessary evil designed to prevent more deaths from a protracted conventional war. That may have been true, but the atomic genie had been let out of the bottle and new, evil rationalisms would find other justifications for its use. Above all, nuclear weapons seemed to confirm our sense that there was something unprecedentedly and uniquely corrupted about our age.

'They are clearly', the English novelist Martin Amis has written of the weapons, 'the worst thing that has ever happened to the planet, and they are mass-produced and inexpensive. In a way, their most extraordinary single characteristic is that they are man-made. They distort all life and subvert all freedoms.

Somehow, they give us no choice. Not a soul on earth wants them, but here they all are.'[10]

In the postwar world, our world, therefore, science was scarred. It was like us, ambiguously corrupted. It was either potentially evil of itself or it led human beings into areas of knowledge we could not control.

It had also created its own strangely twisted moral logic. Albert Einstein had first resisted then supported the development of the atom bomb. His resistance was based on the obvious intuition that such a bomb could only be a new evil in the world. But that was an absolute judgement and, as we know, the great power of infinitely progressing science and technology is to relativize everything. The overwhelming argument for supporting The Bomb that finally changed Einstein's mind was the fear that the enemy might build one first. There was a balance of risk that could only be certainly decided in your favour by the decision to build. The Bomb makes these rules: you can't win and you can't get out of the game.

Such an argument, of course, will apply for ever and to any device, however insane and costly. In the long years of the balance of nuclear terror that only finally seemed to be coming to an end – if a temporary one – in the late eighties, one of the more nightmarish devices with which children used to frighten each other was the Cobalt Bomb. This, it was said, would cause a single explosion that could destroy the planet. It would be used, we argued in the playground, by some crazed dictator backed against the wall. From total war that implicated us all in the battle, we had moved on to the necessary apocalypse, the rational, understandable global suicide. Even to schoolchildren it made perfect sense.

And who can finally say what politician was doing what and why in those strange, still days of the Cuban missile crisis in 1962? I was eleven and perceived it all through the gauze of that age's partial lucidity. And, again perhaps as a function of my age, I imagined that, for the duration of that crisis, all the world was eleven years old.

We saw Kennedy and Khruschev and dimly recalled that they stood for antagonistic systems. But all we really knew for sure was that there was a logic and that it was, in some way, our own creation. We knew it was a dream, but we could not wake up.

Nevertheless, most of the time, nuclear anxiety was containable. Indeed, its very totality gave it a remote quality. Our powerlessness was a real powerlessness that left us a choice only between neurosis and proceeding with our lives as best we could. And, in these lives, science and technology retained much of their problem-solving mystique. Clouds of affluence continued to gather, bringing with them a continuing and accelerating downpour of technological benefits. Meanwhile, Europe and the United States still believed they could feed and rebuild the world on the basis of a few, simple, clear-sighted solutions. It was as if science had taken on the duality of its creators – benign in one aspect, unspeakably malignant in another – and as if our futures all depended on the balance of these aspects as they emerged from the laboratories and the equations.

But even the consolations of affluence were to be tainted in the twentieth century. For affluence also had its problems and ambiguities. It too was as destructive as it was creative. And, with this realization, was born one of the most potent and effective faiths of our day – the faith of the Green.

Concern for the environment is our age's mechanism for resolving the contradictions inherent in the two opposing aspects of science. Environmentalism is based on a scientific insight and yet it is violently opposed to the effects of most of the more obvious and spectacular achievements of science and technology. It is a way of turning science against itself, of rejecting the progressive ideals of economic growth by using scientific means to expose them as potentially suicidal. It is the single most successful popular solution to the terrible contrast between penicillin and atom bombs, air-conditioning and concentration camps. And it is the faith that has intervened between ourselves and the easy superabundance of the supermarket shelves.

This popular success is new, though the underlying belief system is not. The opposition of nature to industrialization began with romanticism in the late eighteenth century. This was an ambiguous beginning. In many ways the worship of nature was as much the legacy of the scientific enlightenment as was the industrial revolution. The romantics' belief in the autonomy of nature was inspired by a sense of awe at its grandeur that paralleled the enthusiasm of the scientists for its vast, impersonal systems.

Yet romanticism did establish the idea of scientific-technological man as the despoiler of the earth. In the later nineteenth century this developed into a more systematic anti-mechanistic view. This was holistic biology, an insistence on the interrelationships of the natural world rather than on the importance of the separate elements. In addition, the idea of energy economics was born. This involved calculations based on the stocks of scarce and non-renewable resources, the working out of a balance sheet of man's consumption. It was a mechanical view, but it did not share the optimism of the mainstream of nineteenth-century materialism. From these two strands of thought, modern ecology emerged.

The importance of this family tree is that it indicates that ecology and environmentalism, its political arm, are not simply objective reactions to a scientific insight. They have a cultural background. This background connects them to all the other attempts I have described to come to terms with science. Equally, Green ideas are not simple notions that can be easily categorized in terms of other political systems. As Anna Bramwell points out, ecology can only really be understood as an entirely new political category. The general point is that greenery is a new twist to an old anxiety.

Finally, there is an important wider dimension which goes some way to explaining the extraordinary fervour and conviction that the modern Green Movement has provoked. The pre-scientific God confirmed man's centrality as well as his dominion over the earth. God had placed man in the world and would look after them both. The widespread displacement of God by science in the nineteenth century also dislocated the relationship between man and the world. Suddenly it became unclear what this relationship ought to be.

The implication of straightforward technological growth was that the relationship was that of master to slave, the earth was available for our exploitation. The opposite conclusion of ecology was that man had become the god of the earth, he must tend and preserve her, he must protect her delicate balance and harmony.

This view produced a steady, though specialized, tradition of thought during the first half of this century. There was a variety of anti-progressive movements providing a pastoral haven for certain types of dissent from the modern world. The politics were variable. Some were socialist, though Marxism itself, because of its industrial and economically progressive emphasis, has always been thoroughly anti-Green, however much its contemporary apologists may pretend otherwise. Some of the ecologically concerned drifted to the far right and, indeed, Fascism and Nazism both employed powerful green rhetoric as part of their dreams of a new, purified race.

But environmentalism did not escape from the intellectual's realm of ideas into the bourgeois realm of popular conviction until well after the Second World War. It was only then that the simple dualism of a benign science helping humanity and a malign one prepared to destroy us all was revealed as far too crude. For, it transpired, even the benign aspect of science produced the most appalling complications. Tracing the precise beginning of modern greenery is not easy. There are many early indicators. But one book, published in the late fifties, can safely be given much of the credit. This was *Silent Spring* by the biologist Rachel Carson.

As I have pointed out, sensitive souls in the nineteenth century had instinctively felt there must be something fundamentally wrong with the soot-blackened, smog-choked cities with their diseases and overcrowding. But the objection was spiritual and ethical and it seemed, in any case, to be progressively weakened by the public health measures that gradually began to clear the slums and suppress disease. Technological greed alone would not, perhaps, destroy us, though the science that controlled it would need to be as robust as that which had created it. The embryonic ecology movement made little impact on this optimistic balancing act within the popular imagination. Things could still be seen to be getting better through the application of benign technology.

What Carson saw and appeared to prove with extraordinarily persuasive clarity was that greed could well be our undoing through far more subtle mechanisms than any we had previously conceived. Her book, in effect, anticipates everything the environmental movement was ever going to say. She knew nothing of the Greenhouse Effect, CFCs or the hole in the ozone layer, the environmental anxieties that obsess us today. But her underlying analysis, morality, even aesthetic, make her the authentic prophet of modern greenery.

In the polemic of *Silent Spring* the greed was primarily agricultural. Carson documented the kind of chemicals that were being used to protect crops and kill insects in the farmlands of the United States. Much of the power of the book arises from the way this innocent, virtuous landscape is revealed as stained and poisoned by the activities of men made barbarians by technology.

The grosser abuses provide one part of the story. At the most horrific level some of the chemicals were so toxic that farmworkers died simply from skin contact. But these were obvious and easily-identifiable crimes, perpetrated by irresponsible companies and farmers. Such things, one might easily say, should be stopped rather in the way that one stops a drunk driving a car. This did not mean there was anything wrong with driving – the use of chemicals – in itself.

The real power of Carson's book and the extraordinary persuasiveness of its legacy lie in the more subtle case histories in which even apparently careful and rational control went disastrously wrong. She writes, for example, of Clear Lake, California. A species of gnat plagued fishermen and holidaymakers. Until 1949 this was just one of the problems of nature, a price extracted for the use of the countryside. But, then, the chlorinated hydrocarbon insecticide DDD, a close relation of the more notorious DDT, became available for disposing of the gnat nuisance. The application of the chemical was done with immense care to ensure that there would only be one part of DDD for every 70 million parts of water in the lake. It worked but, by 1954, the gnats had returned and DDD was applied at the rate of one part per 50 million parts of water. Birds began to die. In 1957 there was another application. More birds died. This time fatty tissues of the western grebe corpses were analysed. They were found to contain 1,600 parts per million of DDD. More animals were examined and it was found that the concentrations of the chemical rose through the food chain from plankton, through plant-eating fishes to carnivorous fishes up to birds. The highest concentration was found in a carnivorous fish – 2,500 parts in a million compared with the 1 part in 50 million of the 1954 application.

The cautious administration of the initial dose had been revealed to be meaningless. The mechanism of the food chain and the way the bird and fish bodies concentrated and retained the chemical meant that, whatever the dose, it could eventually build up to toxic levels. Viewing the lake as a simple dispersive mechanism in which the DDD would merely kill the gnats and vanish like morning mist had proved a ludicrous over-simplification.

The first point about this revelation is that it represents a fundamental challenge to the simplifying instincts of science. As I have described, simplicity has been both a key element of the experimental method and a foundation of the idea of science from the beginning. The success of simplifications and the examination of isolated systems in the quasi-ideal circumstances of the laboratory seemed to indicate that this was, indeed, *the* way of understanding the world. From this it is a short step to assume one can, equally simply, intervene in the world. This chemical kills gnats. We dispense it in the lowest possible dose so that it could kill or harm nothing else and the problem will be solved.

But the miscalculation lies in the belief that the simplicity we require of science can also be found in nature. The plants, gnats, fish, birds and, of course, humans, are caught up in a system of fabulous complexity that defies the elementary mechanism we have tried to impose. Our arrogant simplicity has been humbled by this awesome – but delicate – complexity.

The imaginative force of the insight cannot be overstated. Our simple chemicals had been revealed to be as potent as the atom bomb. They disordered the very heart of nature. As Carson points out, the ecology of Clear Lake, of any lake, field or forest, had been evolved over hundreds of millions of years. It worked because of the operations of deep time. It was robust within the confines of what could possibly happen within the variations and catastrophes wrought by nature. Yet man was not nature – he was, in natural terms, an impossibility – and a few years of chemistry could destroy it all. Suddenly deep time – the faithless abyss into which the biologists and geologists of the nineteenth century had peered – became benign, a kindly sculptor of the delicate pastoral balance. We, with our instant, simple solutions and our short-term obsessions, were the destroyers.

And we were also self-destroyers for DDD could build up in human tissue. We could destroy nature to feed ourselves more efficiently or to stop the gnats biting, but, in doing so, we could poison ourselves. Environmentalism's most effective and persuasive insight is to reveal humanity as rebelling against Mother Nature while still being her dependent offspring. We were fouling the nest, abusing the sanctity of home. The Green Movement has made new Adams and Eves of us – proud, excessively knowledgeable defilers of the earthly paradise.

The aesthetic of Carson's book was pastoral-romantic: the rural idyll of romanticism contrasted with the contaminated hell of industry or the cold simplifications of the laboratory. But its intellectual content added an important and tougher element to the romantic dream of nature. It added the notion of the finite system. Clear Lake was a microcosm of the entire planet. It was complex, but it was enclosed. The chemicals persisted and concentrated to produce unpredictable effects. The same could happen to the planet as a whole. It too was complex enough to defy analysis but enclosed enough to become a prison or killing ground. We were not gods of a universe that our intellects had tamed; we were fragile organisms clinging to a rock that was only precariously habitable.

When the first photographs came back from space of the blue-green, cloud-wrapped earth, Carson's lamentation was shockingly endorsed. Everything that we were or had ever been had happened on this little sphere drifting in the darkness. We were in this together: every human being and Mother Earth.

The climax of this particular theme in ecological thinking was James Lovelock's Gaia hypothesis in which he suggested that the earth as a whole was, in fact, definable as a single organism. It was not a passive system, rather it was capable of reacting so as to preserve its own environmental equilibrium. A complex system of feedback loops sustained a friendly climate for life. It was not that the earth had been an organism before the onset of life, but that life itself had started these loops which had turned the planet into a single organic system. For Lovelock the vision placed man in a position of existential choice, a position beyond all theory in which we had to know ourselves as part of a causal whole from which we could never be untied.

'There can be no prescriptions,' he wrote, 'no set of rules, for living within Gaia. For each of our different actions there are only consequences.'

Such theories have a complex appeal. As existential prescriptions they echo the youthful demand of our day for a programme of action and a system of identity. But they also combine a Christian sense of sin in our defilement of our paradise with a Judaic need for atonement. They are also universal. Whatever else we may disagree about, we can, from the perspective of the environmentalist, surely agree upon the need to stop the planet being destroyed. In this sense we can see that the environment had become one of the first truly global causes. Clearly if the environment is being destroyed everybody is threatened and everybody is implicated. Cutting down a tree in Brazil is no longer an act of merely local significance, it threatens to kill us all. Furthermore it validates the pastoral impulse. From Rousseau and Wordsworth the conviction that the natural is good, the artificial bad has been embedded deep in the industrialized soul. Environmentalism appears to show that this is more than an emotional impulse: it may be true, it may be a matter of life and death.

From the early seventies onwards the systematic, institutional backing for this green truth became overwhelming. In 1972 came the celebrated Club of Rome report entitled *The Limits to Growth* which significantly included in its sub-title the phrase 'predicament of mankind'. The phrase signals the cosmic significance that had already been attached to the Green insight. The key to the report's method was the use of computer modelling and systems analysis – systems theory had grown out of sociology in the fifties and sixties and represented an important attempt to harden up the 'soft' sciences by attempting to make them computable.

The idea of the Club's team was to create a model of the world system which included as many of the relevant variables as possible – population, food production, industrial capital and output, land yield, pollution and so on. It was accepted that this would be approximate and incomplete, but, the compilers believed, it would be accurate enough to show general trends. It would also be flexible enough to allow different policy, technological and accidental assumptions to be built into the computer run. The central assumption of this method was a kind of bureaucratic realization of the imaginative insight granted by the pictures from space – it was the assumption of the finite system, that the earth was a limited resource, adrift in unfriendly and unsustaining space.

The three primary conclusions arising from the computer runs were: the limit to growth would be reached at some stage during the next hundred years and this would probably result in a catastrophic decline in population and industrial capacity; the trends leading to this catastrophe could be altered so as to attain a sustainable condition of global equilibrium; and, finally, work towards this more benign outcome must start as soon as possible to ensure the best chance of success.

The most obvious characteristic of the model's forecasts was the devastating effect of exponential growth. Curves of growth in population, pollution, resource depletion and so on were shown to be steepening uncontrollably. Indeed, so vicious were these curves that they could overwhelm any number of the most optimistic solutions built into the computer runs. We could, for example, enforce worldwide population control tomorrow, but the effect would not be felt for some time and, by then, industrial production, resource depletion and pollution would have engineered the catastrophe. The point repeatedly made in the report was that the hour was late and the most terrible characteristic of exponential systems was their ability to move from total success to total failure in an instant. And, in that context, 'taking no action to solve these problems is equivalent to taking strong action.' The report is repeatedly at pains to stress its compilers' conviction that no merely technological solutions to this predicament can conceivably work.

'The application of technological solutions alone has prolonged the period of population and industrial growth, but it has not removed the ultimate limits to that growth.'[11]

In other words: the dream of an omnipotent, problem-solving technology was over.

The report's proposal was for a system of global equilibrium in which all new development could only be replacement. So new investment in industry, for example, could only take place in line with the depreciation rate of previous investments. The rich would inevitable grow poorer and the poor richer until harmony was attained. Consumption growth would cease, though the report insisted this need not affect 'human growth', an important if impossibly vague sop to the spiritual vanities of the report's liberal compilers.

'The concept of a society in a steady state of economic and ecological equilibrium may appear easy to grasp, although the reality is so distant from our experience as to require a Copernican revolution of the mind.'[12]

The report's position had a simple, mechanical, easily understandable appeal. Anybody can see that if you drink water from a glass, the amount of water in the glass decreases and, finally, there is none. The living cycles of the planet were more complex, but the overriding principle was the same and, in its simplicity, it was far more powerful than any of the fragile feedbacks that sustained our existence. Indeed, the simplicity of this idea was visionary, religious and easily powerful enough to affect the way we buy soap powder. The ecological movement had its evidence. It was as if some strange miracle had occurred and the cult that had been forecasting the event had, at last, been vindicated. And, by an exquisite coincidence, in 1973 came the first 'oil shock' – the major producers increased the price of oil fourfold, sending salutary shivers through the industrialized nations. All that the Club of Rome had been told about diminishing resources, sudden catastrophes and so on seemed to be coming true overnight.

In the twenty years since the report was published, however, it has become clear that its case was overstated. Most of its forecasts have, in the short term, at least proved over-pessimistic. This is not, of course, to say that the principle of exponential growth overcoming the capacities of the planet is fundamentally wrong. Clearly an error of a few decades when we are calculating a catastrophe on this scale is trivial. It may well be that we are facing an ecological Armageddon. On the other hand, we may not.

In effect, this means that the issue of environmentalism itself is largely a question of faith. Do you want or need to believe that we are destroying our own habitat? As with so many issues of faith, the answer is less important than the question. For it is the posing of this question that is the new element in the human imagination.

Behind the question lies the conviction that we are now mechanically capable of destroying the living system of the planet. Clearly there have been apocalyptic prophecies in the past – many expected the world to end in AD 1000 and most universal religions include an arrow of time that flies from a creation to an apocalypse. Sometimes human behaviour is implicated in this end: in the Bible mankind's sinfulness inspires God to flood the earth and most catastrophes, in the Christian world at least, are accompanied by the gloomy certainty that men have brought them upon themselves.

The retributive tone persists in ecological prophecies of the end. But what is new is the strictly mechanical nature of our guilt. To the Green no external force is punishing us for our sins. Rather there is a simple, causal relationship between what we do and what will happen to us. As Lovelock says, for each of our different actions there are only consequences. The glass will empty because we drink all the water. The human race faces extinction because we breach the ozone layer, overproduce greenhouse gases or exhaust our resources. We are bad, dangerous passengers on the spaceship and our folly will also be our punishment.

This sense of our capability for total destruction clearly arises directly from our industrial, technological and scientific success. The scale of what we do is what convinces us. Now we find pesticides in organisms thousands of miles from where they were originally applied and acid rain falls on forests hundreds of miles downwind from the responsible power stations. As with the long-range bomber or the ballistic missile, pollution denies anybody the power to say they are innocent or uninvolved. We *know* we are capable of ending it all, so we know we are all potential victims.

Such knowledge ties the public and the private, the large and the small into a tight moral knot. Nothing can be excluded from the environmental insight – neither the decisions of government nor the impulses of individuals. This is the secret of its power as a quasi-religion. It provides universal values and meaning. It humbles the individual and mocks the scope of human knowledge while still encouraging its devout application in the name of the faith. It also has the capacity for the open-ended demands and the infinity of requirements that characterize any religion based on a view of human inadequacy. To the environmentalist a day can be as full of religious observance as a monk's. He can choose his food to avoid chemicals, factory-farming and blighted origin. He can reject over-elaborate packaging, conscientiously re-use plastic bags and walk or cycle rather than drive. He can proselytize, campaign and demonstrate. In an advanced society there is virtually nothing which he is unable to identify as environmentally damaging. Indeed, to be advanced at all is a form of crime.

For, to the environmentalist, the world is suffused with baleful portents, it is enriched with meaning as in the vision of a saint. It is, above all, *a world*, a unity as opposed to the fragmentary, incomprehensible mass of facts provided by the scientist or the modernist artist.

This vision induces a complex anguish. The achievement of hard, classical science was ambivalent: in one sense it involved the subjugation of nature, but, in another, it asserted our hopeless connection to nature. Science, via its handmaiden technology, showed us how to exploit nature on an unprecedented scale, but it also, most obviously via Darwin, told us that we were part of nature, merely another contingent fact in a contingent world. Clearly the two ideas can work together: we exploit nature because that is the way of nature. The lion eats the gazelle: we build our factories.

Such naturalistic fatalism could effortlessly underwrite the idea of progress. Our very naturalness justified our conquests. In this context the logical position of the environmentalist becomes one that stresses nature's strangeness, her difference from us.

And this is the position of Bill McKibben, in his book *The End of Nature* (1990). He places at the heart of his argument the insistence that nature is actually defined by her separation from the human realm. 'Nature's independence', he wrote, '*is* its meaning; without it there is nothing but us.'[13]

McKibben is a purist, but a reasonably coherent one. For him there is no compromise, no possibility of an ecological armistice.

If nature is defined as everything that is not us, then nature is dead because we have imposed our will and our pollution on all earthly nature. He cannot, for example, enjoy the summer because he knows the climate has undergone artificial change. And this atrocity has been perpetrated very recently and by just a small part of the population: 'the way of life of one part of the world in one half-century is altering every inch and every hour of the globe.'[14] Survival may be possible, but it will scarcely be worth having for we will be obliged to survive in the midst of our own artificial creation rather than with nature.

Finally, McKibben's purism drives him to an important moral extremity. He contemplates the idea of a genetically altered rabbit – 'Why should we have any more reverence, or even affection, for such a rabbit than we would have for a Coke bottle?'[15]

Again the point is not the answer but the question. This question defines the idea of a reverenced or liked rabbit as one that is utterly natural, untouched, in its essential nature, by human hand. A laboratory rabbit or perhaps a rabbit bred as a pet would both fail this test – they would hardly be rabbits at all. It is at this point that the ecological argument is at its most imaginatively persuasive and its most dangerous.

The persuasiveness arises from the pure, radical simplicity of the ecological vision – we must touch nature, if at all, as lightly as possible. For what is touched is ineradicably tainted.

The danger is the fanaticism and absolutism that lie behind this simplicity. McKibben's rabbit may seem like a harmless enough example. But what about a human being who has been genetically altered to prevent transmission of a hereditary disease – an entirely possible technology in the near future? Would he too be beyond reverence or affection? At the more immediate level it is significant that McKibben's pastoral polemic is based on his own affection for the American wilderness. But what of the man-made landscapes of Europe? Nothing in the English Lake District or the Yorkshire Dales has remained untouched by man. Are they also beneath reverence, beyond affection?

In addition, the emotional appeal of the environmental faith in this heightened form produces a certain terrible moral blindness about its effects. It is understandable that after two centuries of triumph for classical science, there should be a reaction against its remedies, innovations and theories. But the reaction can clearly go too far and miss its real target. The pesticide DDT is perhaps the most widely known of the 'evil' substances exposed in the wake of Carson's book. But, by its efficacy against the mosquito, DDT eliminated malaria from large parts of the Mediterranean. Banning it prevented the same thing happening in Africa. As a result malaria still kills and debilitates millions. Are we sure we did the right thing? Are we convinced the moral scales were balanced?

It is clear that the single-mindedness of McKibben's 'deep ecology' springs from a kind of loathing of human intervention of any kind and a mistrust of all human solutions. It is also clear that such an attitude is the logical climax of the environmentalists' position. Dismayed by the visions of destruction and apocalypse, they look for culprits. And they find not just this chemical factory or that power station, they find the human race itself. It is this that drives them to assert the absolute otherness of nature. It is important to note that in much environmental rhetoric – particularly arising from the Gaia hypothesis – there is the gleeful insistence that nature may carry on without us. In polluting the planet we may simply extinguish our own life; other life will continue – cockroaches, anaerobic bacteria or sharks.

This perspective reveals the anarchic and destructive side of environmentalism. Human life, if it continues at all, will, we are told with smug conviction, have to be more primitive. People will have to content themselves with a lower level of affluence. Our entire culture is a wasteful, polluting aberration. Primitive, peasant societies are repeatedly held up as models of human life lived in harmony with nature in salutary contrast to our own exploitative and demanding ways. We – meaning the scientific civilization – are *wrong* in the world, and must change our ways or be evicted.

So environmentalism is a religion of rejection. It confronts and attempts to overcome the imaginative crisis created by the pessimism of the scientific vision by simple denial. Old science was proud and dictatorial. In contrast, the science that cares for the environment approaches nature in humility. Science, after all, had to humble itself in order to understand what went on at Clear Lake or what now seems to be happening to the ozone layer. Science is humbled because in the damaged environment it glimpses the error of its ways. The Cartesian consciousness pushed nature out there, alienating our minds from its processes and convincing us, finally, there there were no values to be found in the world. But the environmental consciousness does find values there. The values are of interdependence, of harmony and balance. They demand sensitivity. Above all, they demand an acceptance of complexity as a positive feature of the world that may perpetually keep the whole truth of nature beyond our rational grasp.

Environmentalism, whatever the truth of its diagnosis, is a cry from the soul of modern man. We want more but we do not know where to find it. Science's effectiveness seems undeniable yet its actual effects on the world are dangerous, disgusting, destructive. So the environmentalist tries to turn the effectiveness to more benign ends. He attempts to soften and control science. He attempts to humanize it by forcing it to co-operate with the organic systems of the planet.

As a substitute for religion, as a metaphysic, however, it is inadequate. Its obvious defect is that it offers only survival. The environmentalist may enthuse about the peace of mind he may attain through correctly green behaviour. But, at base, his reasons for that behaviour are purely practical. There is no transcendant rationale. It is a religion of catastrophe. We can only undo the harm we have done; we can aspire to nothing higher. All that we have achieved is as nothing before the mute, alien spectacle of nature. And that remains, as in the bleak vision of mechanical determinism, all that we can ever have, even in the Green paradise.

6
A new strange mask
for science

It is wrong to think that the task of physics is to find out how nature is. Physics concerns what we can say about nature.

Bohr[1]

The public image of science changed in our century. It changed because the smiling mask it had been wearing suddenly fell away to reveal a face that was as horrible as it was wonderful. Primarily this happened because science over the last hundred years has become so visible to so many. A technological explosion as well as environmental anxiety, nuclear weapons, mechanized total war and all the moral and political complexities of economic growth have put science at the centre of the public realm. It has been brought to trial before a new kind of jury – the jury of popular sentiment, whose verdicts are cruder and whose anxieties more politically potent than those of the philosophers. Suddenly science's achievements can simply be viewed as crimes, its knowledge as sin.

The importance of this for my argument is that it means science has been judged from the outside. The pursuit of objective knowledge for its own sake is no longer the private mission of an elite, subject only to its own demands and sense of virtue. In such an enclosed context it could allow itself to believe its knowledge did include the only truth, that it would one day encompass the entire universe both human and inhuman. But its sudden obvious success both as creator and destroyer convinced us all that science lacked some vital human input. If it could do so much to our world, science could no longer be free. For its very autonomy, which had once been its proud badge of independence from authority, might now be seen as a blank cheque, rashly handed to a greedy and destructive child.

From this perspective, faith in science begins to look like irresponsibility. We had allowed science a dangerous liberty, a removal from the limitations of ordinary human concerns. We had done this in honour of its rigour and effectiveness. And we had done so because, according to the wisdom of the Enlightenment, knowledge must, by its very nature, be free of our subjective values. It was the only way the scientific world could be certain that it *was* knowledge as opposed to just another point of view. Science, indeed, had offered us an escape from the tyranny of mere 'points of view'.

But what the horrors and anxieties of the twentieth century revealed was that such a severance of knowledge from value has terrible implications. Philosophers may have been aware of them for centuries. Now, after Hiroshima, after Dachau, after Cuba, we all are.

All of which is true. But the point is that the scientist himself can – and does – ignore such anxieties. A physicist, chemist or biologist can easily construct a wall in his imagination between his work and the wider issue of science in the world. He will argue, as most do, that human knowledge is a progressive, inevitable and value-free investigation of the nature of the world. It will happen whether we worry about it or not. Certainly he may be aware of real ethical and political problems, but these occur only *after* the facts of the hard science. The atom is 'split' and later comes the moral quandary over the use of nuclear weapons.

This division allows science to retain its authority. In spite of what may happen in the outside world, the scientist can still be convinced he is on the one true path to truth, complete truth. Any shortcomings of science when it is brought into contact with the world arise because the truth is as yet inadequate, incomplete.

And the division between scientific knowledge and the world produces a cast-iron moral defence. The question of whether to employ the atom bomb, the scientist will argue, is precisely the same as the question of whether one uses a gun. A discussion of the moral status of the weapon is irrelevant or meaningless; all that really matters is the soul of whoever might pull the trigger. Equally, the use of a pesticide can be decided upon by a perfectly clear method of balancing risks and benefits that would have been comprehensible to the guardians of our welfare long before such chemicals were created. Nothing has changed except the effectiveness of the tools, the scale of the possible error.

In other words, the inner fabric of science itself could easily remain intact whatever dramas of doubt and mistrust were being staged in the outside world. After all, even the atom bomb was evidence of science's one overwhelming claim on our gratitude and admiration: it works.

But another mask also fell away in the twentieth century – the mask of scientific classicism.

The word 'classicism' means many different things in many different fields. The simplest way of defining it here is to describe it as the view of science that prevailed up to 1900. But this is inadequate since many, perhaps most, scientists retain an essentially classical outlook. So classicism ought best be understood as the view that there is an objective world outside ourselves which is completely accessible to our observation and reason. As material beings we are part of this world, but as reasoning subjects we have the capacity to observe it from outside. Classicism is, in essence, simply the old faith that science is the path to the truth, the whole truth.

The mask of this classicism fell away because the body of knowledge which held it in place was suddenly revealed as radically incomplete and, in large areas, fundamentally wrong. For our age's bright, unstable intensity was not to restrict its gaze to science in its external role. Indeed, what has happened *inside* modern science has proved more extraordinary, more extreme and more confusing than anything that has happened outside. Hiroshima and Clear Lake, California, have had their revolutionary correlatives inside the laboratories and minds of the modern scientists.

I will address the vital issue of whether the old mask that fell away revealed only a deeper form of classicism or something far more strange in this and the following chapter. What is at stake is whether new forms of science offer us a new vision of ourselves in the world to replace the pessimistic visions of classicism. If they do then a revolution has occurred. If science can reveal a new spirituality, then the Enlightenment is at an end. The

Enlightenment said no values or meanings could be found in the mere facts of the world. But if new forms of science prove they can, then all has changed. Science would be no less complete and triumphant, but it would at last have become a fully human form of knowledge.

There are many examples that could be used to show the way science in our age changed. If you are a convinced classicist you will say that these changes represent only an expansion and correction of the existing body of knowledge. If you are not, you will say that they represent something quite new, a new type of knowledge. I will use three very familiar examples to illustrate the nature of the change. They are familiar because they have become emblems of the arcane knowledge of twentieth-century science. As such they all point to a similar movement of the imagination away from the precise but limited ambitions of the nineteenth century towards something more comprehensive and more strange.

The three examples are all called theories: relativity, quantum and chaos. They are often used to mystify and amaze laymen with their bizarre conclusions, observations and implications in the physical world. As usual, however, for my purposes it is their meaning that comes first.

The three theories are fundamentally linked; first, by the way in which they overthrow the old type of mechanical vision and, secondly, by the way in which they potentially transform the idea of scientific truth. Classical science told us that truth could be ours; this new science tells us either that it might be beyond our grasp or, at the very least, that it is infinitely stranger than anything we could previously have imagined. Whichever interpretation you adopt, even the most hardened classicist would agree that 'common sense' will be of no help to your understanding.

But first 'truth' like 'classicism' needs to be examined as it is precisely truth that is at stake in this new science. It is often used simply to mean completeness. The Truth can be synonymous with 'The Whole Truth'.

As I have said, technological success and the vast structure of apparent theoretical consistency had, by the end of the nineteenth century, convinced a significant number of people that human knowledge was within sight of a form of completion. We were about to attain The Whole Truth. It seemed possible that, in areas like biology and physics, the primary contours of this completion were now known. If Newton was right and Darwin was right, all that remained might be regarded as the filling-in of details. There was much still to be known, certainly, but not necessarily anything *fundamentally* new.

As Albert Michelson, America's first Nobel prizewinner, said in 1902, 'the more important fundamental laws and facts of physical science have all been discovered and these are now so firmly established that the possibility of their ever being supplanted on consequences of new discoveries is remote.'[2] The irony of this remark was that it was an experiment of Michelson's, described below, which was to undermine this very certainty.

Such confidence is not unusual in periods of rapidly increasing affluence or, indeed, at any time during the most successful phases of scientific thought. Today it is still quite common to meet physicists prepared to believe that their subject can be brought to a conclusion within a few years or decades at the most – Stephen Hawking being the obvious example. We may dismiss such beliefs as merely crude, ahistorical arrogance which manages to ignore the hundreds of examples of similar certainty that have crumbled under the assault of later orthodoxies. Perhaps, however, it makes more sense to see the confidence as an essential working tool. To ask a scientist genuinely to accept the possibility that, in relative terms, the world will be no nearer the truth at his death than it was at his birth is probably to ask too much. He needs the consoling spectre of ultimate truth if he is to continue his work.

But, to the rest of us, there must always be a general sense in which completion in any area of human knowledge seems inherently impossible. On the evidence of our own lives – our ignorance and insensitivity about ourselves and others, our mistakes, our insecurities – we feel that nothing can ever be known to be final. Our world is uncertain, unstable and subject to constant change; our everyday conclusions are never better than provisional and never can be. Instinctively we know the same must apply to *all* human conclusions. Michelson, Lord Kelvin and the Harvard physics professor with their confidence in the imminent completion of their discipline had their reasons. But they were offending against our intuitions.

In the event, they were also offending against the real condition of science, in particular of physics. For, in spite of their great confidence, terrible fissures had already begun to appear in the clear determinism, the rigid mechanism of the Galilean-Newtonian classical system.

Completion as an idea had long been with us. It was, of course, an essential quality of the cosmology and physics of Aquinas. In that case intellectual consistency was taken to indicate its finality. Aquinas was overthrown, but only by another system which also implied – perhaps demanded – completion. The logic of the Newtonian picture of the universe was as conclusive as that of the Thomist and it had the additional advantage of being more intuitively acceptable to the mechanics of the new age.

Newtonian objects from planets to particles rebounded off one another obligingly and memorably like billiard balls – the most instantly popular image of the system. These objects acted upon each other at a distance by means of forces. These forces were, perhaps, less easily understood. Did they exist as material entities or not? But the intuitive and mathematical power of the whole structure was great enough to overcome any ambiguity in their definition.

Later studies of electricity and magnetism at first presented no problem – their power was found to decline as an inverse square of the distance in precisely the same way as Newton's gravity. Light too could be incorporated either as a collection of particles or as a wave motion in a hypothetical medium known as ether. There were problems about various necessary adjustments and inclusions, but none seemed fatal. The world still made perfect Newtonian sense, or, it could be assumed, would do so in due course. The pattern held.

With Michael Faraday, however, the adjustments began to seem awkwardly contrived. Faraday, born in 1791, was an experimenter of genius. He became interested in developments in the early years of the nineteenth century which had begun to cast doubt on the conviction that all forces, like gravity, depended solely on the distances between objects. Moving electric or magnetic charges were found to create forces, and crucial connections were discovered between electricity and magnetism themselves. With the literal clarity of the experimenter, Faraday examined these phenomena using the now familiar schoolroom tools of magnets and iron filings. Place some iron filings on a piece of paper with a magnet underneath. Vibrate the paper slightly and lines of magnetic force appear. The random clutter of the filings is shaken into order. The pattern of the field was revealed. This field had an actual existence outside the magnets themselves – the behaviour of the filings demonstrated that. And the field persisted in a vacuum.

In the Newtonian system, fields – for example, gravitational or magnetic fields – were not required to have any such physical existence. They were merely a kind of equalizing process that balanced out the accounting of the whole structure. They could be immaterial like the laws of nature. But Faraday, through his experiments, convincingly demonstrated that fields did exist physically and that, somehow, they could propagate themselves through empty space. It was a puzzling revelation that could neither be adapted to a Newtonian model, nor to common sense.

The precise implications of Faraday's experimental findings were realized mathematically by the Scottish physicist and mathematician James Clerk Maxwell. With dazzling insight, he arrived at the equations that would explain the field effects and demonstrate the behaviour of electromagnetic phenomena as they were propagated through space. He even calculated the speed of this propagation and discovered it to be the speed of light. The discovery of this finite velocity of light had been proved, indeed measured, by one Ole Romer in 1676. But it took Maxwell to employ it to unify the whole electromagnetic spectrum. The fact that light did have such a velocity, always and everywhere, was the point. This speed – written as 'c' – was to become the strange, implacable absolute that was to underpin the physics of the twentieth century.

With Faraday and Maxwell something profound had begun to change in the scientific picture of the world. The Newtonian model had described the world as it was and provided an utterly consistent mechanical model. It did not point to any further mysteries. But Maxwell's equations and Faraday's fields did point to a deeper realm, a range of phenomena which were simply not acknowledged by anything within the Newtonian system and which did not behave like billiard balls.

And, I repeat, these new autonomous fields that carried energy from place to place were emphatically not the stuff of common sense. It is important to stand back from our own age fully to understand this point. We know electromagnetic waves well enough today because we can press buttons on boxes and activate a circuitry that plucks invisible waves from the air and turns them into the voices and images of radio and television. We may even be aware that light is one part of the electromagnetic spectrum. From time to time we may awake to the miraculous quality of what is happening, but, habitually, we think nothing of it. Yet this is only habit. Just a little more than a hundred years ago, such an idea would have seemed madness and, more importantly, in precise defiance of all that the wisdom of science had taught us. What meaning can be attached to common sense in view of such changes? With what confidence can we cling to any contemporary certainties about the possible and the impossible? The cold rigidity of classical science was being replaced by something even more terrifying. But what was it?

In the mind of the Dutch physicist Hendrick Antoon Lorentz a further piece of Faraday's and Maxwell's subversive jigsaw fell into place. He established equations of motion for charged particles which, when combined with the Maxwellian equations, provided rules for both particle and field behaviour. More details of the underlying order were being exposed, though, at first, painfully slowly. Nevertheless the implications were ominous and mysterious – the most obvious being that there *was* an underlying order of which we had previously known nothing.

And, finally, there was the most spectacular negative result in the history of scientific experiments. The existence of an all-pervasive substance known as the ether had long been posited as a way of retaining the Newtonian mechanism – Newton himself had speculated about its nature as a means of explaining all actions at a distance like magnetism, gravity or the propagation of light through a vacuum. This ether flowed through all things and spread light waves like ripples on the surface of a pond. Its existence accorded with the mechanical, common-sense view of the world since it permitted non-scientists the possibility of visualizing what was going on, of 'seeing' reality in their imaginations. It potentially saved the billiard balls.

So, for example, it was clear, if such a substance as the ether existed, that the motion of the earth through ether-filled space would cause an 'ether wind' which should produce a variation in the speed of light depending on whether the beam was flowing against or with the wind. In 1887 the Americans Michelson and Edward Morley tested this with a wonderfully ingenious apparatus for reflecting light back and forth over sufficiently long distances for the time elapsed to be measured. They discovered there was absolutely no variation. Either the ether did not exist or there was something very odd about light. Both, in the event, were to become essential truths of the new physics of which Michelson and Morley were to be the baffled midwives.

Such portents, and many others, began to signal an appalling realization: the Newtonian picture was radically incomplete. Its observational and experimental power made it clear that it could not exactly be described as wrong. Its effectiveness seemed to be such a conclusive proof that it was true. But there was clearly something missing, something enormous, perhaps something quite different.

Before looking at the colourful and fantastic spectacle of what was missing, it is necessary to understand the philosophical implications for the idea of truth of the revelation that Newton was, if not wrong, then incomplete. For, in some ways, these implications are as important as the precise details of the error we had uncovered. They point to the heart of our age's crisis of 'truth'.

The first issue is whether incomplete actually means wrong. Touchy and arrogant though he had been in life, Newton himself had too great and, ultimately, humble an imagination to think that he had solved the entire problem. He longed for completion and certainty, but remained, at the last, fully aware that it lay beyond his grasp. His image of himself as a boy on a beach simply picking up a few interesting pebbles indicates that he knew that the scientific project had only really begun with his synthesis.

Yet this synthesis was staggeringly powerful, so powerful that it eventually eliminated all competing forms of knowledge from the Western imagination. Even the alchemy and magic that had lived side by side with science in Newton's mind was overcome by his vision of the celestial game of billiards. Its success engendered a scientific confidence that did indeed begin to believe a Newtonian completion was possible. And the implication of that was that Newton was completely right – the rest of science would be mere filling in of details. This may still mean that Newton's wisdom was incomplete, but only in a very weak sense of the word incomplete. It meant merely that all the implications of his underlying truths had not been calculated.

Incomplete in a strong sense would be more worrying. It would mean that his system was not universally applicable and therefore not true. It would reduce his system to the status of a highly successful model or approximation. Newton would potentially suffer the same fate as Ptolemy. In the light of the idealistic grandeur that had been attached to his system, to prove this was precisely to prove him wrong. *All* was not light simply because God had said: 'Let Newton be!' Most alarmingly for the convictions of the scientists, their effectiveness would have to be separated from 'truth'.

Such considerations raise awkward problems for the understanding of science. If Newton could be wrong, how on earth could we know what was right? The conceptual shock would be the correlative of that which had shaken the foundations of Jesuit wisdom in the early seventeenth century.

In science's first flowering its explanatory and experimental success suggested that here was a method of discovering ultimate truths about a real world that existed beyond ourselves. Things were not the way they were because authority had ordained it, they were thus because that was how nature was and neither reason nor conviction could change that implacable reality. Man was humbled by the impersonality of nature.

But, even in this phase, we have to be careful about the idea of truth. Galileo saw that Jupiter had moons and we might then say that this was a 'truth' of the same order of truthfulness as saying that elephants have tusks. But it is a specific truth. We cannot derive from it the conviction that all planets have moons any more than we can say that all elephants have tusks. Indeed, in both cases, we should be wrong.

A wider truth involves a hypothesis that certain aspects of observation or experiment can be generalized. The behaviour rather than the mere existence of the moons of Jupiter, for example, can be used as evidence of the force of gravity and of uniform motion. Such a hypothesis may lead to a theory and finally a law – in this case Newton's gravitational equation and his laws of motion. In the case of elephants' tusks we might ultimately arrive at natural selection. This is the very essence of classicism in science.

This process has been much discussed, particularly since the radical incompleteness of the Newtonian model was revealed. The whole scientific method has been minutely analysed with no clear result. The key problem is that, if hugely successful theories can be found to be wrong and 'truths' can be found to be false, what can possibly be the real nature of the form of knowledge we call science? Why is it successful and why should we believe it?

I shall return to this (I am destined never to leave it). But, for the moment, it is worth isolating two points about the method of classical science. The first point is the familiar one that it *was* unarguably successful. It worked. This cannot be emphasized enough. In our age – since the new science appeared to undermine classicism – there have been many smart and sophisticated attempts to explain away Western science as merely the product of a particular culture at a particular time. This is an attractive view as it removes at once the problem of scientific truth. In fact, it removes the idea of truth from science completely and replaces it with the concept of the viable or acceptable. It puts science in its place by denying its claim to the highest possible prize.

But all such theories have to cope with the fact that no other local, cultural products have been so universally effective. An experiment performed in Chicago will give the same results if it is performed in Tokyo. Indeed, it will give the same results if performed under precisely similar conditions on the moon or in some distant galaxy. (Some may quarrel with this last location, but the point, at the current state of our knowledge, holds.)

If science is no more than a local cultural product, this universality is the most inexplicable and bewildering coincidence imaginable. It is the equivalent of discovering a Van Gogh on

Mars. It is also a unique state of affairs. Painting or music have utterly different histories in Tokyo and Chicago; religions are mutually contradictory; politics have only in recent decades begun to converge. But there is no Japanese science as distinct from American science. There is only one science and, in time, all cultures bow to its omnipotence and to its refusal to co-exist. The only reasonable conclusion appears to be that, for some reason, science is the one form of human knowledge that genuinely does give us access to a 'real' world.

The second point may go some way to explaining this success. Science, both in the classical sense and in much of its twentieth-century incarnation, is a daring simplification. Galileo's most celebrated experiment – dropping two different weights simultaneously from the top of the Leaning Tower of Pisa – was intended to demonstrate the uniformity of the effect of gravity. And, to Galileo and the enthusiasts watching, it did. Indeed, it still does to us because that is how we are taught to understand the world – as subject to underlying laws. Two objects, whatever they weigh, accelerate at the same rate when in free fall. They would, therefore, hit the ground at the same time. The problem is that the weights did *not* both hit the ground at the same time. Air resistance, a condition local to any experiment conducted within the earth's atmosphere, caused a slight variation. As far as Galileo was concerned this variation did not matter as it was too small to be in any way related to the difference in the weights. Air resistance might, for example, cause a variation of one per cent, whereas the difference in the weights of the two objects might be 50 or 100 per cent. His point was made: difference in weight in no way affected acceleration due to gravity.

Now it would be perfectly possible and perfectly in accord with common sense to take the view that, because the weights did not fall at the same speed, Galileo, who had forecast that they would, was wrong. But, if the world had taken such a view, science would never have happened. Our knowledge would have been stuck in the infinite chaos and complexity of local conditions. Every weight dropped from every tower would be a fundamentally different event. We could never generalize, never make meaningful distinctions, never simplify.

But, in Galileo's imagination and in that of our age, every

falling weight *is* fundamentally the same. It is subject to underlying laws which may then be modified by local conditions. But we can only say anything meaningful or useful about the world by stripping away local complexities to reveal the essential simplicities of these laws.

The impact of this aspect of scientific thinking is as profound as any event in history. It encouraged the human imagination to escape from the bonds of earth and of ourselves. Nobody directly perceives the world as a Newtonian system or in its ideal Galilean simplicity. Indeed, everything in our experience rebels against such a perception – we do not see uniform motion, we cannot employ frictionless surfaces. But, once we accept that all that happens is a local variation on the theme of a network of underlying laws of transcendent simplicity, then we also accept that our experience of the world is profoundly superficial and hopelessly prey to the accidents of local complexity. We have been subjected to the conviction that simple reality lies concealed beneath this complexity.

Classical science, therefore, works because it simplifies. It takes on only those problems that can be solved by the known method. The entire scientific edifice, for all its hermetic inaccessibility to the uninitiated, is a vast monument to simplification.

To bring this back to Newton: simplification can also explain why a theory can be wrong or incomplete but effective. Newton's great simplification was a brilliant synthesis of the behaviour of matter within certain large-scale generalizations. They only break down once the margins of speed, time and space are attained. As far as the general behaviour of matter perceived on the scale of human perception is concerned, he was right. This was supported by his mathematics which, as in the case of fields, could fill in the gaps of theory. Through mathematics a field could be incorporated simply as a way of balancing an equation.

Simplifications, therefore, can be staggeringly effective even if not correct or complete. Some would take this point further and say that simplifications, precisely because they are simplifications and therefore fundamentally artificial, can never be correct or complete. To ignore the complex data of experience is to ignore everything. Others, like the English scientist Peter Atkins, believe the scientific method of simplification is right because, at heart, reality itself is simple. In his hard, pro-science polemic *The Creation*, Atkins writes of taking us on a journey on which he will reveal that 'there is nothing that cannot be understood, that there is nothing that cannot be explained, and that everything is extraordinarily simple.'[3]

This is the hardest of hard classical science, which relies on the apparently common-sense assertion that there is a real world out there which, in time, we can fully understand.

But, I am convinced, the common sense is only apparent. We only find it in Atkins because the last 400 years of intellectual and scientific history have placed it there. For his assertion is, in reality, a faith, a belief that the simplicity we require really is in the world. Why should the world be simple? Who made that decision? Who imposed it? There is no answer, for nowhere can we find any such guarantee. The leap from effectiveness to truth is a leap of faith. The reality of Atkins' assertion is that it is a statement of this faith. And its passion arises from the way his faith has been tested by the revelations of twentieth-century science, most vividly of quantum and chaos theory. For these revelations are, at heart, revelations of complexity. Relativity provides a slightly different kind of revelation.

Back to the story. The ominous legacy of Faraday, Maxwell and Lorentz first began to reveal its true nature in, conveniently enough, 1900. In October of that year Max Planck was walking in the Grunewald woods in Berlin. At the end of that walk he recorded his conviction that 'Today I have made a discovery as important as that of Newton.'[4] He was right.

His discovery was simply a number. It became known as Planck's Constant and is generally written as 'h'. This number expressed an utterly novel insight. Planck had concluded that energy was only ever emitted as a series of 'quanta' or packets. These were very small, but they were, nevertheless, separate.

In fact, the smallness of the quantum was exactly what had ensured the success of the simplifications of Newtonian mechanics. For the Newtonian approximations worked perfectly on the assumption that energy emission was a continuous spectrum. This is a form of common sense, for it is how we live our lives. As the sky darkens at night or lightens at dawn, we do not think of it as darkening or lightening in a series of jumps, we see it as a smooth, continuous movement. Equally, if the air grows colder or warmer we think of it as doing so by passing in a smooth progress through all the intervening temperatures. Planck's 'h', however, showed that this smoothness does not apply at the level of the very small. He had seen that nature was fundamentally discontinuous. The smooth lines we see are, in fact, a series of unbelievably tiny jumps – the celebrated 'quantum leaps'. If these jumps had been big, Newton would never have begun to make sense because the energy spectrum would have been so obviously discontinuous.

It may seem a small, rather specialized point – and, literally, it is a very small point indeed – but the full implications of that insight are shattering, strange and utterly at odds with common sense. In fact, we still do not yet really know what all those implications are. The quantity 'h' is like the solution to a detective story of which we have not been told the plot.

There was even something symbolically magical about the way Planck arrived at the number. He discovered it simply as a way of solving equations rather than via any route through the intuitively possible or the experimentally observable. This evokes the method of that fictional hero of the age of science, Sherlock Holmes, as he affirms it to the long-suffering Dr Watson in *The Sign of Four* in 1889. 'How often', he asks impatiently, 'have I said to you that when you have eliminated the impossible, whatever remains, however improbable, must be the truth.'[5]

However improbable . . . anybody not shocked by quantum mechanics, Niels Bohr was later to say, has not understood it. Erwin Schrödinger was to describe the truths of the new physics as not quite as meaningless as a triangular circle, but much more so than a winged lion. The underlying message of both remarks was that quantum physics could not be made to accord with common sense or intuition. It was bizarre, absurd. Unfortunately it just had to be true, the numbers said so. Newton and Galileo had prepared us for this by showing that the truth lay in universal laws that lay far beyond the limits of our everyday perception. But their versions of those laws still lay well within the range of the intuitive. What was to emerge from quantum theory was to challenge our ability even to guess at the true nature of the world.

There was, the quantum told us, a reality outside the human scale of perception and far beyond the realm of human intuition.

This is the insight that returns us to the dizzying sense that our perspective can never be quite right. It is the insight that we are always destined to be the wrong size. As Swift's Gulliver or Carroll's Alice expands and contracts, as Dante finds himself stuck in an inconclusive middle way, as every philosopher since Descartes has wondered whether we are men or gods, so quantum theory exposes to us the ambiguity of our position in the universe. Newtonian 'truth' was true for people-sized perceptions. If we were the size of atoms, it would be a comical distortion. We could not even dream of the billiard balls, only of limitless fields of indeterminacy.

The revolutionary strangeness of this reality has meant that we still have a long way to go before we can understand its meaning, even if we restrict ourselves to the realm of physics. As for its implications for our imaginative grasp of the world of our daily lives, these have hardly begun to surface. Some have found God in the quantum; others simply a vast area for new exploration in physics. Some say its discovery signals, yet again, the approaching end of human knowledge about the behaviour of matter; others that it reveals only the still unfathomable depths of our ignorance.

But, before describing some of the developments of quantum theory, I shall discuss a parallel revolution that occurred five years after the Grunewald walk and which, perversely, was to contradict the quantum insight, thereby creating an abyss in our knowledge of physics which has still to be bridged.

In 1905 Albert Einstein published four papers. All four were revolutionary, but the third – *On the Electrodynamics of Moving Bodies* contained the Special Theory of Relativity. Ten years later Einstein produced his General Theory and his revolution was complete. Newton was overthrown.

Einstein abandoned the Newtonian fetters of absolute time and space. The two were relativized and unified. They were aspects of the same underlying reality. Space and time formed a continuum that curved and enfolded about itself. Gravity was a distortion of this continuum caused by the presence of mass. Mass and energy were correlated according to the formula e (energy) = m (mass) × c². c was the magical constant in the whole system – the ultimate speed, the absolute velocity at which any signal can be propagated. It was the one absolute in the relative universe of Einstein and it creates a ghostly cage around all that we can know.

These generalities are now familiar enough, but their consequences remain as bizarre as ever. Objects shrink when they are in motion, spacetime curves, light is bent by gravity and so on. Yet such phenomena occur in conditions so remote from our experience that we still live in a world that is, to all intents and purposes, Newtonian.

Again we confront the contradictions of scale. Quantum theory revealed that our intuitive perceptions and the mechanical extensions of them were wrong at the level of the very small. Relativity revealed they were also wrong at the level of the very large. The world we see is a function of our size. The truth of which we had been so proud was exposed as no more than *our* truth, a local simplification, a way of talking about things.

'It is wrong', said Niels Bohr, one of the greatest architects of quantum theory, 'to think that the task of physics is to find out how nature *is*. Physics concerns what we can *say* about nature.'

This was the supreme statement of the new anti-classical faith. Science was not the Royal Road to the Truth, it was a kind of ambiguous compromise with nature.

Yet, for all the revolutionary impact of Einsteinian physics and for all the radical developments in physics that took place both because of him and in his lifetime, the most important thing to note about the man himself is his determined classicism. In a profound sense he believed he had done nothing fundamentally new. He had not overthrown Newton, he had merely built upon his achievement.

Einstein was a classicist almost to the point of obsession. In essence this means that he saw the foundations of the scientific project as still intact. By means of reason, experiment and observation the human mind was capable of establishing the nature of a world beyond itself. This was a real world which behaved according to classical notions of causality and consistency. Every effect had a cause that could potentially be ascertained and the same physical laws applied throughout the universe. Einstein overthrew one form of classicism by creating another. He discarded the classicism that had become corrupted and earthbound by its entanglement with the idea of common sense. He replaced it with something far more powerful, strange and flexible. But he passionately retained the central classical doctrine that the world was accessible to human reason. It was perhaps this combination of his rational, humanistic passion combined with the strangeness of his insights that have made him the emblematic and popularly sympathetic figure of modern science and the very icon of the contemporary idea of genius.

This new classicism was reinforced – as had been the old – by the grandeur and accuracy of its predictions. One of the few experiments that could be carried out to ascertain the validity of the General Theory was to establish whether light from stars beyond the sun was deflected by the sun's gravitational field. This requires an eclipse. The astronomer Arthur Eddington exploited an eclipse in September 1919 to observe the appropriate stars and, sure enough, they appeared displaced almost exactly in line with Einstein's equations. In November of that year the results were announced to a joint meeting of the Royal Society and the Royal Astronomical Society. The mathematician Alfred North Whitehead described the occasion in revealing terms: 'There was a dramatic quality in the very staging – the traditional ceremonial, and in the background the picture of Newton to remind us that the greatest of scientific generalizations was now, after more than two centuries, to receive its first modification.'[6]

It was the correlative of the predicted return of Halley's comet on the basis of Newtonian mechanics. The comet did return; light did bend; the universe obeyed the equations. It worked.

In spite of his classicism, outside the strict realm of physics Einstein could be wildly inconsistent. For example, although he believed in the possibility of expressing everything scientifically, he acknowledged that such an expression would be meaningless – 'as if you described a Beethoven symphony as a variation of wave pressure'.[7] But to say it would be meaningless is exactly to admit that you cannot, in fact, express everything – the meaning would be missing.

In such a context, invoking God, as he so often does, may be seen as rhetorical rather than literal. It is a way of expressing the fundamental and universal nature of Einstein's project. But the importance of the invocation is that it indicates Einstein's deepest feelings about the classicism of the task of science. It is to explain everything, to attain the ultimate generalization, the supreme simplification.

The tragedy of his life was that, after 1915, this approach led him into an increasingly sterile desert of speculation. He devoted his days to establishing a 'unified field theory' which would finally bring together the laws of physics in a system as complete as that of Newton. He failed, either because he did not know enough, specifically in the subatomic field, or because such a theory is not possible. But more painful than any cause was the way he failed. For he could never bring himself to accept the strange and uncontrollable body of theory that was to spring from Planck's quantum. Much of the drama of his later life arises from his celebrated but doomed attempts to show that the very quantum theory which had formed such a significant part of his own special relativity was, in fact, inadequate.

The unity which Einstein sought is still being sought today – it is the unity of the Theory of Everything. But Einstein had been pursuing a unity based upon the final denial of the strange inconsistencies that emerged from quantum theory. He thought these inconsistencies arose simply because we did not know enough. Once we had the right equations the more bizarre, counter-intuitive aspects of quantum theory would vanish. Today most agree the unity can only be achieved by finding some way of making quantum theory and relativity work together. For it has been the strange destiny of twentieth-century science to have evolved two theories of immense power – some say they are the most powerful ever – only to discover that they appear to be contradictory.

The quantum world that appears once we wave the magic wand of Planck's 'h' is the world of the inconceivably small as opposed to the vast planetary and celestial movements involved in relativity. It is the world of the physics of matter rather than cosmology. Perhaps, therefore, we should not be surprised if the theories we require to explain the very large and the very small are not in accordance. Unfortunately, it is essential to the scientific imagination that, in some way we cannot yet grasp, they must agree. A contradiction cannot be built into nature, it can only be in our minds.

The discovery that energy was emitted in discrete quanta and the ensuing body of quantum theory – the main body of which is known as quantum mechanics – are in practical, earthbound terms probably the most successful single scientific innovations of all time. All our electronics are based on quantum theory's overpoweringly accurate forecasts. If practical success establishes truth then quantum theory is as 'true' as anything we have ever known.

The problem is that it appears to overthrow the entire classical ideal. It does so in a number of ways.

First Planck's quantum indicates there are only certain possible energy states. Since an energy state determines the orbit of an electron around the nucleus of an atom, this means that only certain orbits are possible. Nature cannot sub-divide 'h'. So when an electron acquires enough energy it will move from one orbit to another. But it does not 'move' at all. It simply disappears from one orbit and reappears in the next. In between there is nothing. It is as if we had decided to pick up a drink and seen it move from the table to our lips without passing through the intervening space. But the quantum of action is so small we do not see the effect. As when watching a film we do not see the individual frames, only the illusion of movement. The quantum truth is that nature is, at heart, a series of stills.

This affront to common sense goes further. Radiation occurs because of the emission of particles. Statistically we may know with immense precision how many particles will be emitted in a given time. But, at the quantum level, it transpires that we can know almost nothing. How many, yes, but why this particle as opposed any other, no. This may at first seem harmless. But consider the implications. What quantum theory is saying is that, however detailed our observations, we cannot ever know which particle will be emitted next. This means that we cannot discover a cause for the emission. Quantum theory appears to show that there is a limit to causality – inconceivable in any fully classical system.

In the extraordinary and instinctively anti-classical mind of the quantum theorist Werner Heisenberg the mystery deepened yet further and began to include ourselves. The quantum was not just a physical limit, it was also an epistemological limit, a frontier beyond which our knowledge could not pass. Built into the fabric of our world was the astonishing certainty of uncertainty. We could not know everything about a particle. The better we knew its velocity, the more vague was our knowledge of its position the better we know its position, the more inadequate was our knowledge of its velocity. This was not a problem of observation, it was a calculable consequence of experiment, theory and observation. It was not simply that *we* could not know these facts, it was rather that *they could not be known*

Worse still, the fact of our observation appeared to affect the real world. Particles changed their nature according to whether they were being observed. Just as later philosophers adapted Descartes's 'I think' to become 'There is thinking going on', so quantum theorists have effectively changed the experimental 'I observe' to 'There is an observation going on'. The subjective and the objective are blurred in the quantum world.

Meanwhile, Bohr established his celebrated, indeed notorious, principle of complementarity. This said that a single conceptual model may not be enough to explain all observations of atomic or subatomic behaviour in different experiments. The most familiar example is the wave versus particle complementarity in the behaviour of light. Sometimes light behaves as if it consists of particles, tiny objects flying about the place. Sometimes it acts like a wave, a ripple as on the surface of a pond. This duality is absolute. When light is a wave it is unarguably a wave; when it is a particle, there are no means available for denying its particleness. Bohr's complementarity worked simply by accepting the duality. Again classicism is offended – there must be some single truth. But, for Bohr, physics was only about what we could *say* of the world, not what we could know of the final truth.

The challenge to classicism was on a bewildering number of fronts. If events like radiation were not subject to causality, did this mean that, at heart, our world was acausal? If there was a fixed limit to our knowledge of matter, did this mean that the entire classical project would have to remain incomplete? If the observer affected the observed, did it make sense to talk in classical terms of a world beyond and independent of our presence? If reality behaved differently in different circumstances, did this mean there was no reality?

Quantum theory even had a specific booby trap for those who may have thought that at least Aristotle had been ejected from respectable science. For it turns out he was right all along: quantum theory establishes that a vacuum cannot happen in nature. A field cannot be given both a precise value and a precise rate of change; quantum fluctuations lead to the formation of pairs of 'virtual' particles like protons or antiprotons. We have to say space is full of virtual particles that can spring into existence at any moment. Far from being empty, space is a plenitude of potential. Indeed, that is the only meaningful way of defining space. A vacuum cannot exist because existence exists.

A hole appeared to have been torn in the fabric of classical science and many concluded that the entire project must be at an end. The austere, classically inclined Planck was horrified at what he had done. In 1933 he felt obliged to insist on the pre-eminence of the principle of causality: 'because scientific thought is identical with causal thought, so much so that the last goal of every science is the full and complete application of the causal principle to the object of research.'[8] Science, he was reasserting, *is* causality; his 'h' was just a more precise value placed upon the details of that causality, not some key to a magically unpredictable realm beyond science.

Less anguished but more determined, Einstein commented caustically on the apparent lack of anything more than statistical evidence about the primary constituents of matter by again evoking God – 'He does not play dice.'[9]

God and the ideal of classicism were inseparable. He is the final Cartesian guarantee that our reason does not deceive us. But again Einstein is betraying an important inconsistency. The reason the remark about God and his dice was made and endures in the popular imagination is that it succinctly evokes the need for a *moral* consistency in the universe. It would seem immoral for God to gamble with our destinies like some Las Vegas card-sharp. So Einstein chooses not to believe it.

But what are the implications of that choice? He is saying reality has an obligation to be ordered, a moral obligation. But this is covert anti-classicism. It implies that there is a moral value in nature and this is precisely what classical science denies. Einstein's classicism breaks down because of his emotional need for knowledge and value to be interconnected. He could not believe that the good God made a universe based on this terrible, unreasonable game of chance. Or rather, since his God was more rhetoric than reality, he could not believe that such a universe could be made.

But why not? What if He did play dice? After all, one of the few ways in which quantum mechanics does accord with common sense is that it endorses our everyday sense that the world is a risky and unpredictable place. All that quantum theory appeared to be saying was that it was *fundamentally* risky and unpredictable.

Unfortunately, Einstein's repeated attempts to prove the inadequacy of quantum theory always failed. The most celebrated was the Einstein, Podolsky and Rosen 'thought' experiment in which he attempted to prove that uncertainty was no more than a disturbance caused by measurement. Two particles interact and are then separated by a great distance, no further interactions could apparently take place. But, Einstein pointed out, if quantum mechanics were true, then the particles would still appear to be connected, and idea he found intrinsically disturbing. For one of quantum theory's most bizarre twists is the principle of non-locality. Particles can apparently have effects on other particles at a distance and apparently by communicating with each other instantaneously – in other words, at greater than the speed of light. If we are to retain the speed of light as an absolute, as we must, then the only solution is that in some way far beyond our current conceptions of space and time, they remain the same particle. In recent years this difficult concept has been experimentally confirmed.

So quantum theory was possessed of a rude, insensitive, if eccentric, health which made it impervious to the delicate dismay of the classicists. It worked. It worked not only practically but also theoretically. Most importantly it re-established the possibility of the physical existence of the world after the modifications of Newtonian mechanics had begun to turn even this into a puzzle. At the atomic level, for example, classical theory had been found to suffer from serious deficiencies. Most obviously, a strictly Newtonian view of the behaviour of matter on this scale was not possible as the mechanics would produce instability. The electrons would spiral into the nucleus. Nothing solid could ever exist, as atoms could never come into existence. The persistence and stability of matter had been the real subversive mystery lurking beneath the edifice of Newtonian confidence. Bohr pointed out that the stability of matter was 'a pure miracle when considered from the standpoint of classical physics'.[10]

But quantum theory replaced spiralling electrons with discrete energy states. Atoms could only exist in certain ways and were prevented simply from wobbling in and out of an infinity of possible conditions. Planck's Constant placed a lock on the system. Matter was possible.

In offending against a common sense, quantum mechanics had found a way for our world to be real. Such a solution made the mere fact that quantum theory was crazy seem a feeble irrelevance. It 'saved the appearances' so triumphantly that it must be true. Unfortunately the radical nature of the theory even undermined this kind of inference. Pre-eminently it was Bohr who took the most extreme view of the implications of the overthrow of classicism. For him reality was almost infinitely far removed from all our models, classical or quantum. Our familiar, modern image of the atom, for example, as a miniature planetary system of electrons orbiting round a nucleus was no more than an image, a convenient fiction arising from our calculations. To demand that such a thing exists is a kind of naïvety based on the narrowness of our perceptions. Beneath all appearance, beneath all imagined appearance was only the seething, unimaginable quantum flux and this could not meaningfully be said to 'look like' anything.

Before turning, in my next chapter, to some of the wider implications of all this, there is one further, though quite separate, development in twentieth-century science which reinforces the radically anti-classical message of quantum theory. This is chaos theory. It is an infinitely simpler insight but it shares with the quantum and, indeed, relativity many of the same revolutionary implications. Its most familiar emblem is the 'butterfly effect'. Chaos theory implies that the fluttering of a butterfly's wings in

Tokyo can precipitate a chain of effects that will produce a storm in Chicago.

Simplification is, as we have seen, an extraordinarily powerful tool. It clears the road towards scientific understanding. As I have said, however, it does not necessarily provide a 'true' picture in the sense of one that accords with our everyday experience. Classical science might respond by arguing that the picture was only 'untrue' to the extent that it was inadequately, as it were, classicized. In the case of Galileo's falling weights, for example, we could perfectly well draw the experiment fully into the classical fold by adding equations relating to air pressure to those accounting for gravitational acceleration. Greater degrees of complexity simply demanded larger numbers of classical equations. The apparent complexity of the world was merely a multitude of superimposed simplicities.

Quantum theory threatened this view by questioning the possibility of any such determinism. Imagine a box filled with a gas. The classicist would say that, given all the information about the position and energy state of every particle within the box, he could forecast all its future conditions. The quantum theorists would reply that such knowledge was impossible. The best the observer could hope for would be a statistical likelihood. Complexity is real, it is not just an excess of simplicities.

The chaologists mounted a similar assault, but from the point of view of maths rather than physics. Chaos enthusiasts explicitly define the theory as another, perhaps the conclusive, weapon against the classical system – 'Relativity eliminated the Newtonian illusion of absolute space and time; quantum theory eliminated the Newtonian dream of a controllable measurement process; and chaos eliminates the Laplacean fantasy of deterministic predictability.'[11] There is a danger in such triumphalism but it does express the scale of the revolution.

To start, as always, from Newton: his laws are assumed to be so powerful that any system's approximate behaviour can be calculated on the basis of a rough knowledge of its initial conditions. Planetary orbits may wobble a bit, but, given an object of earth mass moving at earth's velocity around the sun, we can impute the rest. In essence, this says that the ideal, experimental conditions dominate local perturbations. The great modification achieved by Einstein does not affect the principle: in his relative universe, it is still the general laws that reign.

The first indications that this might be inadequate as a purely mathematical model of reality had been identified by Henri Poincaré in the early years of the century. But the most celebrated anecdote about the start of chaos theory concerns a laboratory accident in America in 1961, in which a computer calculating weather patterns produced some bizarre, indeed chaotic results. The cause was a shortcut taken by the scientist. He had fed back some numbers into the computer up to only three instead of six decimal places. A classically inclined mind would have said that the variation should not have been enough to cause significant disturbances in the forecasts. But they produced huge and wildly unforeseeable fluctuations.

This could be taken as a mere oddity. Unfortunately as Lorenz, the scientist in question, investigated further he discovered something utterly fundamental. All systems, simple or complex, are prone to this kind of chaotic disorder and all systems are, therefore, potentially unknowable.

The roots of this idea can, perhaps, be traced to the first applications of statistics. Statistics are a kind of acknowledgement of ignorance. They may be used, for example, to study populations. We may say that ten per cent of a particular population will have red hair. What we cannot say is which ten per cent will have red hair. One might argue that this in itself suggests a breakdown of classical causality. But the obvious response is that it only breaks down because of the amount of information that would be required to make it work. With enough computing power and enough genetic and biochemical information we could indeed forecast redheadedness.

Nevertheless, in all sorts of areas we have to be content with statistics. They clearly represent a different kind of knowledge from mathematical certainty in that they represent a lower, more generalized form of information. But they work as a means of controlling otherwise uncontrollable quantities of material.

The weather provides a much more flexible example. Weather forecasting may logically begin as a kind of statistical art. Summers in the past have always enjoyed average temperatures in a given range so it is reasonable to assume they will in the future, although we cannot say for certain that this particular summer will fall within that range. Still less can we say that it will be exactly 27.6 degrees centigrade in our garden tomorrow.

But, as observational information about weather patterns begins to be available, the classical-mechanistic mind will assume it can do better. One such was possessed by Lewis Fry Richardson who suggested the model for future meteorology in his work *Weather Prediction by Numerical Process*. He estimated that 64,000 people equipped with desk calculators could predict the weather at the speed at which it was actually happening. In essence, Richardson's is the model for modern meteorology. Weather forecasters cast a grid over the surface of the earth and upwards into the atmosphere. Like Mercator's maps of 400 years ago, these throw a conceptual net over the unknown. Readings are taken at points on the grid and these are used to impute the condition of other points. Finally a computer produces a complete, seamless image of the condition of the entire atmosphere. Greater accuracy and greater ability to see into the future, the classicist would argue, will be provided by increasing the number of grid points and increasing the computer power.

The chaologists would disagree. Chaos says that the world conceived simply as mounting pile of numbers is unknowable. The weather forecasters will find themselves subject to a law of very rapidly diminishing returns. Double the computer power may add a day to the accuracy of the forecasts, doubling it again perhaps only an hour or so on. This is more than simply a problem of investment in computer storage capacity, it is a revelation about the nature of our world. Reducing the world to a deterministic, fully causally related system of equations is not possible because complexity will always overwhelm any system. It is not that our computers are not big enough, it is rather that no computer could ever be. A computer constructed of all the energy and matter of the universe since time began would still be found wanting. Certain problems are simply, in the jargon, 'transcomputable'.

The implications for mathematics of this relatively simple, indeed rather innocent, insight may well prove as devastating as Planck's quantum did for physics. Chaos is, in essence, a reversal of the classical view that underlying laws are what count and local perturbations are relatively trivial. In chaos local perturbations can be overwhelming and, just as important, you cannot tell when they are about to overwhelm you. On a weather map, you cannot know which irregularity might explode into a storm.

This even has implications for geometry. Consider a map of a country. We wish to measure the length of the border of this country. We could do so by measuring the border on the map and scaling up to full size. As we measure we follow the lines of beaches, inlets, river estuaries and so on and we arrive at an acceptable approximation which could be used to sail, fly or walk around this country. But suppose we tried to measure this border in the real world. We may walk it or, attempting to be more precise, we may measure all the variations we find. Clearly on this scale a pebble on a beach is larger than the entire beach was on the map. Which is the truer measurement: the approximation or the one that takes in every detail? And at what scale do we abandon our precision: the scale of a grain of sand? Clearly a mile on the map could turn into hundreds of miles if we drew our lines around the contours of every grain. In chaos even the maps of science that once told us 'You Are Here' have lost their conviction.

With fractal geometry the study of chaos has arrived at a system of mapping that, in some way, approximates to the infinite graininess of the world of our senses. Instead of the purity of Euclidean triangles or even the surfaces of topology, these new forms have a generative life of their own. Run through a computer they produce staggering, unpredictable images of caves, cliffs and valleys. Again the idea of the map becomes a symbolic embodiment of the condition of our knowledge.

The high visual content in chaology had proved central to its power. Mathematicians have been inspired by watching, for example, the chaotic flow of water rather than simply by working with equations and have been made to feel that maths has, via chaos, been returned to the real world.

But chaos has also revealed a strange and previously concealed order. Run through myriad variations on a computer comparatively simple equations seem at first to be behaving randomly – producing solutions dotted about the screen. But, over time, patterns emerge. The most celebrated is the bulbous gingerbread man known as the Mandelbrot set, whose complexity is such that every magnification reveals more elaborate layers of pattern and yet who provides a sympathetically humanoid overall appearance. The gingerbread man is an analogue of reality in that he appears to suggest a version of infinity. He is not analogous in the simplified sense in which an equation of classical physics models some aspect of the world. Rather he is analogous in that he mimics the stupendous fecundity of reality, its apparent need for complexity to sustain its existence. He is not simple. He looks like us and he is as indecipherable as we are.

The field of chaos is still in a process of rapid development. Recently at Cornell University a paper – by Christopher Moore – emerged suggesting that we had barely scratched the surface. Beyond the conventional chaos of the 'butterfly effect' lay 'complex chaos'. In the butterfly effect there at least remains the quasi-classical implication that there are initial conditions which could, theoretically, be employed to predict final outcomes. But complex chaos appears to show that this is yet a further illusion. Complex systems, even if fully known down to the level of the butterfly's wings, may remain utterly unpredictable.

We really could not say in any meaningful sense, ever, that Galileo's weights simultaneously hit the Pisan ground.

Relativity, quantum theory and chaos reveal the style of our new science. As the nineteenth century ended in a mood of sublime confidence that human knowledge was nearing completion and our power, through the application of that knowledge, was approaching that of the gods, so the twentieth century began – and has continued – by destroying the foundations of that confidence. Extraordinarily that process of destruction has taken place both from outside science and from within.

What, overwhelmingly, these new theories reveal is the limitations of our old perspective. We had thought the world, understood through the medium of people-sized rational beings, was, indeed, *the* world. Now we know that this was no more than a generalization, a simplification, that worked. We can move in a number of opposing directions from this realization. We can reevaluate all knowledge pessimistically as eternally incomplete and inaccurate or optimistically as infinitely more human than classical science had previously allowed; or we can reaffirm that science is still on the right track, that these vast modern explosions of scientific awareness are celebrations of the heights that the imagination of the Enlightenment has scaled and of the new challenges to come.

Put simply the question is: is science still the bleak messenger or is it now ready to tell us something else? Does all this exotica prove that Darwin and Freud were wrong, that their pessimism about man's ability scientifically to find spiritual peace was misplaced? Or does it simply show that science had grown more complicated, baffling and remote from our need for truth and peace? Is all this, in short, just a new, strange mask for the same old face?

7

New wonders . . .
new meanings?

*It is true that few unscientific people have this particular type of
religious experience. Our poets do not write about it; our artists do
not try to portray this remarkable thing. I don't know why. Is no
one inspired by our present picture of the universe? The value of
science remains unsung by singers; you are reduced to sharing not a
song or poem, but an evening lecture about it. This is not yet a
scientific age.*

Feynman[1]

The new developments I described in the last chapter represented, as I explained, an assault on classicism. That classicism
had reached a climax in the vast pessimism of accidental man
alone in a meaningless, valueless universe. So surely the hope
must now be that new science can lead us away from that pessimism, that, in some way, it may be pointing us to a universal
vision replete with meaning and significance.

Apart from being one of the creators of quantum electrodynamics and, therefore, a key figure in recent physics, the late Richard
Feynman was an emotional man. He seemed, at times, overcome
by the grandeur of science and bewildered by the notion that it
could in any way reduce human experience. The knowledge of
science, he wrote, 'only adds to the excitement and mystery
and awe of a flower. It only adds. I don't understand how it
subtracts.'[2]

Feynman's emotions had, in an important sense, blinded him.
It is, as I have shown, all too obvious how science can subtract
from excitement and mystery. It can subtract everything.
Equally, to say that this is not a scientific age is to take only the
narrow view that we seldom explicitly celebrate science. In

reality, it is all-pervasive, the reluctantly acknowledged 'truth' of our day. We celebrate it by being who we are and doing what we do.

Elsewhere, Feynman tells a story about the astronomer Arthur Eddington, the man who, in 1919, provided experimental confirmation of Einstein's relativity by observing the curvature of starlight round the sun. Eddington one day realized that the stars derived their power from a nuclear fusion reaction in which hydrogen burns into helium. He was sitting the next night on a bench with his girlfriend.

'Look how pretty the stars shine!' she said. 'Yes,' he replied, 'and right now, I'm the only man in the world who knows *how* they shine.'

'He was describing', Feynman commented, 'a kind of wonderful loneliness you have when you make a discovery.'[3]

The story may be apocryphal, but that does not matter.

Typically Feynman acknowledges only the romance of knowledge. It must have seemed that way to the brilliant master of quantum electrodynamics, the arcane mysteries of the interaction of light and matter, as well as to Eddington the astronomer. But perhaps the girlfriend would have seen only mechanical disillusion in the revelation. The stars, her boyfriend had shown, were no more wondrous than an internal combustion engine or an electric kettle. They just burn hydrogen into helium and we are supposed to be impressed. We are supposed to be *grateful*.

But the romanticism of Feynman, his scientific religiosity, should not be lightly dismissed for it raises the difficult question of what it is that we need in addition to the bare facts, however exotic. For him this need is satisfied by the miracles revealed by science itself. It is the usual approach of the popularizers, although coming from Feynman it sounds less patronizing. There seems in him really to be an element of anguish that people do not share his wonder.

But, of course, this approach represents an attempt to turn science into something more than the word normally allows by insisting on an inherent poetry. The trick is to annex the word 'poetry' on behalf of science. The response is to insist that poetry cannot, by definition, be science and science is not enough.

Or is it? Perhaps now it is ready to be enough.

This chapter is about the effects on the imagination and the mind of modern man of the possibility that science can reveal more than grim, mechanistic necessity.

First I shall deal with the imaginative developments. For the sense of something strange and new happening in science has penetrated our culture even down to its most popular and familiar levels. For the first and most direct way in which the new science has revived cosmic speculation from its mechanistic slumbers is in our dreaming, popular mind. In the human imagination, science has once again been allied with wonder. Feynman would have found consolation if he had looked to the movies rather than to songs or poetry.

Science and wonder have always had an uneasy relationship. Clearly the apparent success of the classical scientific view by the end of the nineteenth century had tended to reduce the poetic dimension of the quest. The scalpel had been taken to the stars and our souls were next on the operating table.

Indeed, the success of the mechanistic view almost literally threatened the existence of poetry. As I have said, the Modernist Movement in the arts – which lasted roughly from 1880 to 1930, the years of both technological plenty and spiritual disillusionment – was a concentration on the way things are expressed rather than on what is expressed and, in its most pessimistic form, this emphasis arose because of the suspicion that there was nothing left to express.

Science had robbed us of the myths, metaphysics and illusions necessary for art. T. S. Eliot's poem *The Waste Land* (1922), perhaps the most familiar and exemplary of all modernist works of art, is about constructing cultural forms in the absence of truth. The Hindu evocation of peace with which it ends – 'Shantih shantih shantih' – is ironic. All that this peace amounts to is a bitter and conclusive sterility. The power of the poem springs from its manipulation of the old meanings lying strewn about the culture like the rubble of a bombed building. It did not offer new meanings, new buildings.

Wonder was further and more visibly expelled by the scientific loss of innocence which I have described. Lethal pesticides and atom bombs signalled that, far from being wondrous, our knowledge was little better than demonic ingenuity.

In popular form, however, technology did much to sustain the note of wonder against the tide of dismay. The explosive, modern growth of technological applications produced the optimistic sense of an infinity of possibilities. Infinity is the foundation and inspiration of all wonder precisely because it is what our own mortality most painfully denies us.

Technology offers the imagination a form of release. Similarly 'weird science' like quantum mechanics or chaos theory offers escape. They imply that common sense is inadequate and, therefore, life might really be much more exciting than it appears. Technology realizes the implications of science in the realm of the understandable. The new physics suggests those implications may help us escape from the boringly feasible.

And the two combine to create a kind of magic of consumption: we may know nothing of quantum mathematics, but we can turn on a transistor radio or a compact disc player. This suggests the possibility of a kind of sublimation. We can cope with the cold otherness of science by humanizing it into products and opportunities. And this need not be real. Once the principle is understood we can create myths of technology, fictions to endow science with value, and reinvent wonder.

Before the advent of modern technology and the new physics, science fiction as a literary genre may be said to have provided a negative imaginative response to the inroads of science. Jonathan Swift's *Gulliver's Travels* of 1726 and Mary Shelley's *Frankenstein* of 1818 are both warnings of the moral and social dangers of the new form of knowledge.

But modern sci-fi, with its celebration of the multiple shocks of the new, was born in the midst of the technological explosion of the late nineteenth and early twentieth centuries. Jules Verne and H. G. Wells travelled space and time and fought aliens in anticipation of the new landscape of adventure of which the human race appeared to be on the threshold.

The genre has remained intact and energetic ever since. Indeed, at the level of the strip cartoon, it might be said to be the primary genre of pop entertainment. Batman, Superman and their kin are science fiction – Batman is a master of technology and Superman an alien of strange, spiritual grandeur. Equally, from the explicitly sci-fi *Star Wars* to more oblique fantasies like *Weird Science*, films have leaned heavily on the possibility of technological transformation or alien spirituality.

The creative energy of science fiction derives from its freedom – anything can happen within fictional conventions that can be infinitely loosened by an imaginary technology – and from its ability to revitalize certain myths. *Star Wars* (1977), for example, was a cowboy film set in outer space. *Close Encounters of the Third Kind* (1977) and *ET* (1982) were fables about the rediscovery of innocence. Batman and Superman are both Robin Hoods, strong men on the side of virtue and so on. Science was like a new costume for the old longings of the culture.

From such imaginative impulses one can see the need to redeem science and technology. In *Star Wars* the robots are given fussy, eccentric, lovable personalities. In *Close Encounters* and *ET* the superior technologies of the aliens are shown to be possessed of a benign and wholesome magic denied to the crude mechanisms of earth. In bleak, pessimistic sci-fi – the film *Alien*, for example – a range of anxiety-related myths is reborn and, in this particular case, the clean power of the machinery is revealed as a healthy counterpoint to the slimy, organic threat of the alien. In the sequel, *Aliens*, the heroine fights the monster wearing a huge, mechanized extension of her body. The prosthetic machinery is savage and industrial in appearance in contrast to that in *Star Wars*, but it is virtuous nevertheless.

Both the alien and the imaginary machine represent a wonderous and redemptive challenge to the underlying determinism of our knowledge. Science may strip meaning away from poetry, but it grants opportunity to the fantasist. And it does so with increased vigour with each technological development in the real world as well as with each speculative development in the minds of our physicists. Verne and Wells, for example, were inspired by the explosive technological energy in their world to invent the basics of modern sci-fi: time travel, space travel and alien invaders. And all the strange mystifications in the television series *Star Trek* arise from an awareness of the 'strangeness' of contemporary science. We could not travel with Captain Kirk at 'warp speed' in a purely Newtonian universe.

But the imaginary machines of Victorian and Edwardian fiction are seen as essentially plausible, mechanical developments in an age when machines were likely to be comprehensible in lay terms. The steam-engine was impressive, but it was also decipherable. The further imaginative development comes when machines and theory in the real world take on something of the inexplicable qualities of sci-fi. So, for example, an average person may understand a motor car without too much difficulty. Even a basic computer may make a certain sense. But, once electronics seriously took off, lay understanding was left behind. Few lay people are likely to have more than a dim understanding of even an electronic calculator, let alone the more sophisticated developments that have sprung from silicon and gallium arsenide technology.

At this point the machine becomes an indecipherable black box which, in addition, is likely to be cheap enough to be replaced rather than repaired. Its mechanism is something we have neither need nor competence to explore – the machine becomes as irreducible, as absolute, as a natural object . . . or as ourselves. It is no longer something to which we have to relate in the way we may have done in the past to cars. Rather it is like a rock or a plant, a part of the natural environment which we pick up in passing and discard when it has expired. It has no interior with which we need to concern ourselves.

And, also like natural objects, this allows these machines to take on the lineaments of magic. Max Weber wrote of the way the importance of preaching in Christianity increases as that of the magical aspects of the faith decline. So Protestantism, the least magical form of Christianity, places the most emphasis on preaching. Preaching can be seen as no more than explanation – as a revealing of the mechanism. If we lose interest in the mechanism, as our ignorance obliges us to do in the case of these black boxes, then magic is reborn. The home computer becomes as mysteriously significant as any tribal fetish.

This, of course, expands the realm of the possible. Real sci-fi surprises are all about us. It is a commonplace of contemporary memory that people used to marvel at the first computer games that followed the advent of the silicon chip. A simple tennis game with two white lines for bats and a flashing white square for a ball provoked astonishment. Anybody with the elements of a mechanical education would speculate about how they worked. Within years the effort was abandoned. The games had become so elaborate that mechanical wonder became too plainly inadequate a response. We could never know how such things work; we could only consume them as if our imaginations were irreparably severed from those of the makers of the machines. Yet even they were probably doing no more than assemble black boxes. The complete knowledge of the science and the technology involved is pushed ever further away from the reality of the object itself, concealed ever deeper in its workings.

Along with this is reborn the myth of innocence – our innocence when confronted with the mysteries of aliens and machines. In *ET* it is the child who can relate to the advanced technology of the alien and in *Weird Science* it is computer-literate adolescents who employ electronics to build their fantasy woman. Similarly, in *War Games* it is a child with his computer who is able to 'hack' into the United States defence system and almost precipitate a nuclear war. The image of the teenage American boy alone in his bedroom with his computer and partaking of rites of which his parents have no understanding is as much a part of modern movie vernacular as was the man on his horse in the thirties.

Yet it would be meaningless in such a film for the director or writer to spend much time telling us about the details of the computer itself. We know all we need to know: it is a mass-produced machine with the various add-ons we are accustomed to see. We know we could go to the local computer store and buy the same hardware. Wells' time-travelling machine, in contrast, was strictly a one-off. What counts in a new sci-fi is the interaction between this neutral, indeed quasi-natural, material and the psychology of the boy in question. The implication of *Weird Science* was that a woman-making machine was, in effect, an off-the-shelf product. And, of course, that technology could be employed as a problem-solving mechanism for adolescent traumas.

Much of this method of imaginatively encompassing science is to be understood through the hardware/software duality that plays such a large part in our current popular understanding of the way machines work. A machine in the classical-mechanical sense is dedicated to a particular end and all that is required is that it be started for it to pursue that end. So we start a car and it moves along the road or we wind up a watch and it tells the time. Such machines are determined in every aspect. All that they are is tuned to a single function and, because of that, we cannot ask a car to tell us the time.

But computers – and, potentially, all electronic machines – are different. They are a neutral bundle of circuitry. Left to themselves they do nothing. They are hardware requiring software, a set of instructions, to tell them what to do. And this software can tell them a number of things. The same hardware can process words, draw graphics or make calculations depending on what software we use to instruct it.

The rapid spread of computers into households and offices over the past decade has made this duality familiar. Every child and most adults in any advanced nation know the difference between hardware and software. This explains why, in contemporary sci-fi, the machines themselves are no longer so important. We know the hardware in the bedroom is of no particular interest in itself. The software – both in the child's head and on his floppy discs – is what matters.

At one level this duality clearly has a Cartesian ring. Descartes's mind-body split becomes a software-hardware split. The functional neutrality of the hardware endorses the neutrality of our bodies and the isolation of our minds/selves/souls/software within them. According to this view, machines have finally begun to be like us – random circuitry animated by a programme.

In creating limitless technological possibilities, we could, therefore, have put an impenetrable barrier around our own potential. As in the trivial truism often employed to belittle computers we are no more than the totality of our inputs. This fraught and difficult issue of whether we are machines or can be replicated by machines will be dealt with in the next chapter.

Something more, however, than a revived Cartesianism is clearly at work here. For these ubiquitous, cheap and incomprehensible machines do not simply appear to be leading us to a mechanical utopia. This may once have been the sci-fi dream, but now we require more. Perhaps this is because so many utopias have been seen to fail. Or perhaps it is because mechanics has been understood to be an inadequate foundation for paradise. Possibly even the discovery by modern science of unexpected layers of complexity in the world has opened new imaginative possibilities.

Whatever the explanation the end of technology as an understandable process in the modern imagination has frequently inspired a rebirth of mysticism.

To stay with sci-fi for the moment: the film *2001: A Space Odyssey* (1968) appeared at the height of optimism about our ability to explore space. This was a year before the first manned landing on the moon and it was widely felt that space would soon be conquered by an inevitable process of technological development. The very fact that the moon landing was the precisely timetabled outcome of a pledge made nine years before by President Kennedy emphasized that space exploration was, above all, a problem of the will. The science would happen because we willed it. We discovered problems; science solved them.

Again this has the effect of neutralizing the glamour of the machinery. It drops into the background. The issue of how we shall get there recedes to be replaced by what we shall do and find when we finally arrive. *2001* concerns a trip to Jupiter to investigate a mystery. The hardware is certainly celebrated, but in a significantly mannered way. A Viennese waltz plays over shots of space stations and docking ships. There is a certain effortlessness. This part is easy, the director, Stanley Kubrick, is telling us; what we are to find on Jupiter is the real difficulty.

What we find is a sign of an apparently divine intervention in history. The sign is shown to link all of our past from the first weapon-using ape to the space stations. The sign is a black slab, perfectly smooth and impossible as a natural object in a mechanically conceived nature. It only takes one such object for all our knowledge to be humbled. And, in fact, the film ends with a foetus, man reborn, surveying the blue-green home planet from space. We are being led to the truth. Science is the essential tool, a necessary part of our evolution. But the truth itself is far beyond anything dreamt of by our science. In terms of intellectual history this is the modern version of liberal theology – of a Hegelian world spirit that is leading mankind to his apotheosis.

Science is redeemed by its part in the historical pattern of salvation. Wonder is regained.

Obviously such fantasies can be dismissed as wishful thinking. But they convey two important messages about our age: first, technology is being assimilated and downgraded. It is a natural phenomenon that performs a secondary, hardware function. Whatever we are on the verge of, it is no longer a mechanical Utopia. The effort of technology itself is no longer an imaginative effort, in fact it is almost a bore. In the comic sci-fi novels of Douglas Adams the robot Marvin is afflicted, in spite of his stupendous intelligence, with nothing more than cosmic boredom and depression. Knowing and being able to calculate everything reveals to him only how futile it all is.

The second message is that this assimilation of technology is allowing or forcing us to look beyond. We can no longer derive inspiration from what we can do; what may bring us back to life, however, is the modern discovery that our existence and that of the universe is far more complex and indecipherable that we ever realized. We have the hardware, what we need is the software.

In all of this the culture is telling us that the scientific dream has a new life far beyond the old mechanical reveries. It is a form of the dream that makes Feynman sound dated. Certainly the wonder of science as contemplation of the mechanism is still with us and with our scientists. But these dreams are of something more, of a new completion of vision – not simply that science is wonderful, but that the world is too.

This brings me to the deeper level at which new science has been taken to offer spiritual hope as well as popular wonder. I apologize in advance for the complexity of the remainder of this chapter. I cover a good deal of rather strange ground. I think this is necessary to give the flavour of what I take to be very important developments. These are people's answers to the problems I have been posing. I do not agree with them, but I feel obliged to report them in order to clear the ground for my own answer.

For some thinkers, relativity, quantum mechanics and chaos have opened a window into the closed, darkened, hermetically sealed room of science and even suggested a software solution, a new completion, a new spirituality.

It is not difficult to understand why. Relativity replaced billiard balls bouncing off each other in absolute time and space with the curved space-time continuum; quantum mechanics revealed uncertainty and a limit to causality at the heart of matter; chaos showed there were whole realms of the real world that lay beyond our ability to predict and therefore control. All of them showed that the deterministic, mechanistic science of the nineteenth century was wrong. And, as I said, if that was wrong, might not its accompanying bleak, pessimistic atheism be wrong as well?

The hard science view is that pessimism, bleak or otherwise, is none of its business. This is simply to investigate the real world, whatever the consequences.

'I don't feel depressed about it,' said the zoologist Richard Dawkins in conversation with me, 'but if somebody does, that's their problem. Maybe the logic is deeply pessimistic, the universe is bleak, cold and empty. But so what?'

Some people, even some scientists, however, do get depressed about it and have attempted to derive hope from the fact that the scientific universe of the twentieth century may not be bleak, cold and empty. The point here is not simply that the new science is strange and apparently anti-classical, but also that it has inspired science to ask new questions that border dangerously on the metaphysical, if not the moral. For having spent most of its history assiduously searching for answers to the questions what? and how?, science, even hard science, is now quite clearly in the business of why?

A brief digression which is, I hope, illustrative of this point. I interviewed the physicist Stephen Hawking just before the publication in Britain of his book that gives a popular explanation of some of his thinking, *A Brief History of Time*.

He was understandably abrupt given that his disability – motor neurone disease – obliges him to speak via a computer and voice synthesizer at the rate of ten words a minute. I grasped most of the science and the style of his thinking – strictly hard and deterministic – but it did not seem to provide enough clues to the man. So, a week later, I interviewed his wife. She was distraught and, even in the formality of this first and only meeting, began to pour out her doubts about the direction he was taking. She was a devout Anglican, but he cared nothing for religion. This did not matter, she explained, as long as he adhered to the Big Bang theory of creation. A single event at the start of time clearly left room for God. But Hawking had moved on to his 'no-boundary condition' view which implied no beginning and no end. Instead there was simply an infinity of expansion and contraction as on the surface of a sphere. God was unnecessary.

From her point of view this new form of cosmology could neither be isolated in his mind nor in the confines of an observatory. Its ambitions had grown too vast and too fundamental. He was providing answers to questions which she regarded him as unqualified to ask. They may not have been the answers she wished to hear, but they were answers nevertheless and they were answers to questions which had hitherto been the province of faith rather than physics.

The turning point for the new science, the birth of these wider ambitions, was Einstein's sterile search for a unified field theory. He wanted to explain everything, in effect to produce a set of equations potentially capable of predicting that you would be reading this now. Newton would have considered this neither necessary nor possible. He described the mechanics of a universe held in place by God – the why?, in other words, was God. In addition, the original Newtonian universe was full of things other than physics, including the magical. But in Einstein the magical has evaporated and God only clings on to a weak, rhetorical existence. He is still just about there at the end of *A Brief History of Time* where Hawking writes of knowing the mind of God. But, by now, the deity has become the merest whisper of an afterthought, a functionless grace-note to round off the theory. The true reality for these men, in spite of any disclaimers, lies in the equations.

Carl Sagan puts it succinctly in his introduction to the book: 'Hawking embarks on a quest to answer Einstein's famous question about whether God had any choice in creating the universe. Hawking is attempting, as he explicitly states, to understand the mind of God. And this makes all the more unexpected the conclusion of the effort, at least so far: a universe with no edge in space, no beginning or end in time, and nothing for a Creator to do.'[4]

Such an ambition for absolute completion and an utterly autonomous vision of existence has arisen from the developments of quantum theory and relativity and, most importantly, from the attempts to unify the two. In effect, these theories changed the known universe from a machine into a cosmic system of fields and energy. The stability of matter which we know and upon which our existence depends was explained, but only by the acceptance of an instability at the heart of all matter. Our world was built from a teeming fluidity of arbitrary events, the quantum flux.

As I have said, Einstein the classicist had found himself in the ambiguous position of one of the primary dreamers of this very unclassical dream. He opposed quantum mechanics as a conclusive theory. But he went even further with his own equations. He demanded stability of the universe, even though his own theories seemed to tell him that it could not be. So, in order to sustain the idea of a stable universe, he invented the celebrated cosmological constant. This was the precise force necessary to counteract the effects of gravity which would otherwise, according to relativity, be ultimately terminal, concluding the universe with a final, implosive crunch, just as strict relativity would suggest it had started with a single Big Bang. This cosmological constant was Einstein's real god – a simple number that would preserve the classical balance and do away with bangs and crunches.

But this god also died. The observational evidence that the universe at large was wildly unstable and irredeemably unclassical was overwhelming and has become an essential part of our contemporary imaginative grasp of the cosmos. NASA's X-ray missions in the sixties and seventies revealed a universe consisting, as Freeman Dyson has written, of 'collapsed objects and cataclysmic violence'[5] that utterly disposed of the absolute peace of Aristotle's celestial perfection. And, most important of all, the Big Bang seemed to be true.

Predicted by pure relativity but denied by the cosmological constant, the Big Bang is the explosion that began space and time. There is no 'before' the Big Bang and no 'elsewhere'. It is one limit of classical knowledge; the other being the Big Crunch towards which the universe of pure relativity is heading.

Yet anti-Big-Bang stability theories persisted in competition until, in 1965, Arno Penzias and Robert Wilson in America stumbled upon a radiation field apparently distributed throughout the entire universe. Its characteristics exactly matched those predicted for a radiation fall-out from the Big Bang. This was, in effect, the faint echo to the first ears that could hear it of the beginning of time.

With the Big Bang, developments in quantum theory begin to converge with those in relativity. For it is in that unimaginable instant of violence that the very large is crushed into the very small. The two are forcibly united.

It is also there that we find the epistemological transformation of modern physics. Classicism, to reiterate, portrayed a universe that was simply, mechanically *there* in absolute time and space. Of this thing the question why? could not directly be asked. Indeed, the entire effort of classical science and the Enlightenment had been to establish this division between the why? and the what? Values and meaning are not to be found *in* the world. It ground mechanically on. God or whatever was simply outside.

But the new physics subverted the picture of the universe required by this view. In the Big Bang and related theories it became clear that our knowledge may well allow us to ask why? of the universe. This, in essence, is what is now happening in physics. Quantum theory and relativity have been extended to ever greater extremes to the point where many physicists now feel confident enough to say they understand all physical processes back to the first fragment of time immediately after the Bang. In this they have been provided with experimental evidence from the vast particle accelerators in Europe and the United States which, by creating increasingly extreme conditions, have disassembled matter to reveal both its internal structure and its possible behaviour in the first nanoseconds of time's existence.

The fabulously exotic growths that these theories and experiments cultivate bloom from time to time in popular form. They intimidate and confound. We have, we discover, plunged inwards into spaces far smaller than the atom into a world of quarks and leptons and, outwards, we appear to have travelled 15,000 billion light-years to the edge of the spreading light cone whose point rests upon the instant of the Big Bang. Our equations point us to strange metamorphoses of time into space and space back into time. They point also to an underlying multidimensional reality. A few years ago the 'supergravity' theory pointed to eleven dimensions in which all the things we call forces were, in fact, the workings of the unseen dimensions. This was superseded by a ten-dimensional theory which predicted the existence of superstrings.

Superstrings are the current stars of speculative physics. Think of the size of the whole visible universe. Imagine the ratio of that size to the size of the earth. Then imagine the size of the earth to a single atomic nucleus. The ratios are roughly the same: 10^{20} to 1. A superstring is smaller than a nucleus by the same ratio. These inconceivably tiny entities might, it is thought, be the ultimate constituents of matter, knotting and looping to make time, space and our souls.

Such theories – known in general as field theories – imply a development from an initially unified condition of 'total symmetry'. All fields were one before the Bang; there was no positive and no negative, for all was cancelled out in perfect harmony. This condition is what we might understand – if such understanding were possible – by the word 'nothing'. Everything that exists or happens – all that is, in other words, not nothing – represents part of the process whereby this initial perfection has collapsed. We exist, to the modern physicist, because of countless broken symmetries all of which point us back to an initial symmetrical nothing. We are a blot on the pure void which is the only natural condition, though it is not, of course, a condition since it is not anything.

The name given to these powerful hypotheses that unify forces, particles and laws is 'gauge theories'. They represent an entirely new, total view of matter. As John Barrow has written: 'Gauge theories show that physicists need not be content to possess theories that are perfectly accurate in their description of *how* particles move and interact. They can know something of *why* those particles exist and *why* they interact in the manner seen.'[6]

And Barrow also points out that the way such theories move towards a unifying principle, a Theory of Everything, represents a kind of ultimate expression of the primary scientific belief. This belief he characterizes as the conviction that reality really and ultimately is 'algorithmically compressible'.

This connects the new physics to the classical scientific quest in that both can be seen to be in search of simplifications. Newton was also in the business of algorithmic compression. In this sense, the only difference is the ambition of the simplification – now we aspire to simplify everything where once fragmentary insights seemed to be enough. But Newton would not have dreamed of attempting algorithmically to compress God. We do.

Hard scientists would say that such a distinction between the old and the new physics is irrelevant. The two are continuous; science has never wavered and, above all, science is the only truth we can have. They remain mechanists. The fabric of cause and effect – even if modified beyond the recognition of any layman – can be retained. In this view the success of the new physics does not signal the overthrow of classicism, rather it signals its ultimate affirmation.

One obvious objection to this might be that hard science may indeed describe and explain everything since time and space began, but that merely pushes the question why? to a further extreme. Why, specifically, did time and space begin? Why was the symmetry broken? God has not been ejected, we have simply become more aware of the splendour and complexity of his creation. One response of the hard scientist is to construct pre-Big-Bang scenarios, pictures of the field we call nothing, and speculations about an infinity of other universes. Their point is that there can be no end of our powers of explanation, no end of theory. And this limitlessness does not show that we need God, it shows rather that we are capable of anything. At any stage we will be able to say our knowledge is incomplete, but such incompleteness is not fatal to the structure of science, nor does it re-admit God and the whole multicoloured pageant of metaphysics and theology.

This is, of course, a hopelessly unresolvable argument that depends on attitude and faith rather than persuasion. Either scientists *want* to be mechanistic determinists or they do not. Most do. But there are a number who have challenged the idea. I will now deal with what they say.

As we have penetrated the forces and fields that hold matter and ourselves together, it has become clear that we appear to be living in the midst of a coincidence of quite staggering proportions. If any of the forces and interactions we have detected were different by the minutest amount, then we could not possibly exist. From the moment of the Big Bang, matter appears to have been involved in the most elaborate, finely tuned conspiracy to produce air-breathing, carbon-based life forms possessed of self-consciousness.

The details of this conspiracy are such that we cannot simply dismiss them as pure chance. However, we can change our perspective and realize they we may be misusing the word 'coincidence'. In the thirties the British scientist Paul Dirac discovered – while, it is worth knowing, he was on honeymoon – what appeared to be the most bizarre and extravagant coincidence. In all that we could see through our telescopes and with our minds, we could calculate that there were 10^{78} particles in the observable universe. Meanwhile, the ratio of the strengths of electromagnetic to gravitational forces between two protons is almost 10^{39} – in other words the exact square root of the number of particles in the known universe. It was a staggering connection between the impossibly big and the impossibly small. It is sometimes known, with a strange, innocent charm, as the Large Number Coincidence.

But, as time passes and the universe expands, more light reaches us for the first time. The number of particles increases. The only way the ratio could be maintained would be if the proton ratio also changed. If it did not, this really would be just a coincidence that would disappear as the particle number increased. But that coincidence was so extraordinary that it would be easier to believe the proton ratio did change – even though it was a hard-won constant of nature.

After much experimental work had been done, the American physicist Robert Dicke pointed out in 1964 that this was not a coincidence at all. It had to happen. The ratio existed because the universe must have been in place for that length of time to create conscious observers. It was an 'anthropic selection effect'. It is there because we are here and we are here because it is there. The perspective from which you view the world will determine the truths that you derive.

Dicke's insight is now known as the 'Weak Anthropic Principle'. In essence, this states that we cannot observe the universe from outside, we can only see it from our point of view – that is to say at this particular stage in its development. This is anti-classical because classicism is based precisely on the idea that, in some way, we can observe the universe objectively from outside.

(As a slight digression it is worth noting that this may also be a powerful circumstantial argument against the existence of alien life forms. Physical and biological theories of the way we have evolved point to an immense span of time for carbon – widely regarded as the only probable basis for life – to go through the processes that would finally produce life. Time, in the new physics, also means expansion from the point of the Big Bang. So, though we may look at the size of the universe and think there *must* be other beings somewhere in all that immensity, in fact, it has to be that size just to support things like us. There could, in other words, be aliens but the simple immensity of space is not evidence that there must be.)

This is the formal statement of the weak principle: 'The observed values of all physical and cosmological quantities are not equally probable, but they take on values restricted by the requirement that there exist sites where carbon-based life can evolve and by the requirement that the Universe be old enough for life to have already done so.'

With typical clairvoyance Pascal first glimpsed the anthropic principle when he mused on the oddness of the human perspective: 'Why is my knowledge limited? Why my stature? Why my life to one hundred years rather than to a thousand? What reason has nature had for giving me such, and for choosing this number rather than another in the infinity of those from which there is now more reason to choose one than another, trying nothing else?'[7]

In its weak form the anthropic principle is reasonably uncontentious, but important nevertheless. All it says is that scientists must be on their guard against theoretical or experimental 'selection effects' which might bias their results – results, for example, they might be tempted to generalize, but, in fact, only occur now when we are here. I can explain such bias in terms of the Fridge Light Hypothesis. Every time we open the fridge door there is a light on inside, so we speculate that there is always a light on inside. In reality, opening the door turns on the light – what we see is determined by our presence and our method of observation. In this sense, the anthropic principle amounts to no more than a sound methodology.

But this is not quite satisfactory. We still have the problem that our current understanding indicates the universe could perfectly well have developed in any number of alternative ways. Robert Dicke saw that Dirac's 'coincidence' was actually a fact of life. But that insight did not restore classical harmony. We are still obliged to acknowledge that there is something very odd about the way the universe appears to have slotted itself together to produce beings who could think about how odd it was.

So, perhaps, from the initial conditions a whole range of universes sprang up in which every permutation of physical development took place. Very few, perhaps only ours, would attain the alignment of qualities that would produce life. And yet ours is the only one we can study. The fact of our existence must, therefore, condition the entire history of the universe; it is precisely our presence that distinguishes it from other possible universes. That it is the sort of universe that produces things like us must be our primary insight.

This leads to the Strong Anthropic Principle: 'The Universe must have those properties which allow life to develop within it at some stage in its history.' It must have because we are here and all our studies of the universe are obliged to take this into account.

At this point the anthropic principle becomes very contentious indeed. For to take this view is to regard life as a goal towards which this universe has been heading. Instead of asking questions of the Big Bang along the lines of 'What happened?', we would start to ask questions like 'How did this lead to us?' We are really present in the cataclysmic chemistry of that moment. This is a profoundly anti-classical suggestion which threatens to restore Aristotelian causality by emphasizing the end product as part of the process of causation.

And, as if that were not enough, there is the Final Anthropic Principle. This suggests that not only *must* conscious life come into existence in this universe, but also that, once it has done so, it can never die. The argument is utterly speculative and hugely complex. It involves vast periods of time in which earth's resources are exhausted, we colonize space, the universe begins to contract, at a critical moment in this contraction effectively infinite amounts of energy from 'gravity shear' become available, recolonization of previously exhausted regions begins and, finally, life, or rather consciousness, expands to fill the entire universe.

'At the instant the Omega Point is reached,' explain John Barrow and Frank Tipler, 'life will have gained control of *all* matter and forces not only in a single universe, but in all universes whose existence is logically possible; life will have spread into *all* spatial regions in all universes which could logically exist, and will have stored an infinite amount of information including *all* bits of knowledge which it is logically possible to know. And this is the end.'[8]

One can say little about the validity of such speculations except that they are there and they do much to demonstrate the new style of the scientific imagination. Both the idea of Total Symmetry and the Omega Point suggest a visionary unity and completion springing from within the confines of science itself. These things are not being discussed by theologians or philosophers, they are being discussed by scientists. The book in which Barrow and Tipler ponder the Final Anthropic Principle is a mass of equations. This tells us that the scientific language is aspiring to a far greater universality than ever before. For it is impossible *not* to draw conclusions about the meaning and morality of our lives if we accept anything like the Final Anthropic Principle. Impossible, in other words, to retain the Enlightenment's value-fact division.

Indeed, we do not even have to go that far to grasp the enormity of what is being suggested. The Weak Anthropic Principle alone represents a kind of revolution. When Copernicus first put the sun at the centre of the universe, he was saying that the human race does not occupy a privileged place and all classical science since has been based upon that view. But, as the case of Dirac's ratio shows, such a view produces distortions in our knowledge. In reality, if science is to attain anything like accuracy, it must place human consciousness at the centre of the universe. Classical science aspired to see the universe objectively – that is from a point outside human consciousness and therefore from a godlike position outside the universe – but there is no such position. It was an illusion, a dream.

Such a realization produces an explosion of possibilities in every area of science. If all that we have been doing has merely been an effective series of extrapolations on a series of assumptions which we now know to be flawed, then perhaps the truth of the world is far more radically different from anything of which we have yet allowed ourselves to dream.

The British scientist Rupert Sheldrake has, for example, put forward the theory that our entire view of evolution and causality is wrong. We think of one thing happening and then another and all of this being guided by laws which somehow persist eternally within nature. The past simply provides a causal platform for the present; it has no organic involvement.

Sheldrake finds this improbable on a number of levels and, in any case, these laws of nature are items of faith which can nowhere actually be found in nature. We know of Newton's Laws of Motion but where are they and what are they made of? Instead he proposes that we are made by habit. The past creates patterns that persist via 'morphogenetic fields' which impinge on the present. Memory persists in all things. When one human being learns to ride a bicycle it becomes easier for the next human being to learn and so on. The original effort of learning creates a pattern in a field in which riding a bike is added to the realm of possibility and action.

The idea of a form of memory persisting in all matter and events through the medium of an as yet undetected system of fields is utterly at odds with all previous science. Yet it solves or, rather, bypasses with almost extravagant neatness a number of problems in biology relating to the way organisms are formed. Mechanistically we think we are formed by our genes. But, in reality, we still have no idea how this happens. What our school books tell us is a straightforward chemical process is, in fact, full of gaping holes. We do not know how genes are expressed as organisms either because we are ignorant or, as Sheldrake argues, because it does not happen. Conventional biology will say we merely have a problem understanding the way organic molecules 'fold' themselves into particular proteins. This problem will be solved. But, for Sheldrake, the fact that the problem has proved so intractable is because 'folding' occurs not because of some chemical process we have not yet grasped, but because the molecules are subject to the memory of previous foldings. We are formed by the pressure of the past transmitted through morphogenetic fields.

If this appears bizarre and improbable, he points out, consider quantum mechanics. Within a few years this winged lion had taken over physics. And yet biology persists with its ancient mechanistic thinking. Is it so improbable that a quantum revolution is about to occur in the life sciences?

If the theory is true, Sheldrake concludes, 'we shall sooner or later have to give up many of our old habits of thought and adopt new ones: habits that are better adapted to life in a world that is living in the presence of the past – and is also living in the presence of the future, and open to continuing creation.'[9]

At this point it becomes clear that such new scientific thinking is explicitly and urgently spilling over into new moral attitudes. This does not necessarily happen in the outer reaches of the new physics. If we look into the minds of contemporary physicists and mathematicians we may be startled by the ambition of what they are attempting on the basis of their radical new knowledge, but we may still dismiss what they are doing as well within the bounds of what scientists normally do. The Omega Point may have moral implications, but none are immediate. They do not demand action.

But Sheldrake wants his theory of what has been wrong with our classical knowledge to change everything here and now. It is a scientific way of returning us to a more harmonious position within our culture. Instead of contingent observers of a cold universe, we become creatures made by and making its history. The cosmos embraces us with its fields.

There are many others for whom the restriction of the new scientific knowledge to the minds of scientists is not enough. Quantum theory in particular seems, to some, to demand a reassessment of what we do and who we are. Indeed, the strangeness of the quantum might be the precursor of a new revelation. It may be that the pioneers of quantum mechanics were John the Baptists to a scientific Christ.

'One may suggest here,' David Bohm has written, 'that we are in a position which is in certain ways similar to where Galileo stood when he began his inquiries. A great deal of work has been done showing the inadequacy of old ideas which merely permit a range of new facts to be fitted mathematically (comparable to what was done by Copernicus, Kepler and others) but we have not yet freed ourselves thoroughly from the old order of thinking, using language and observing. We have thus yet to perceive a new order.'[10]

Bohm, an important physicist and quantum theoretician, has been one of the leaders of the attempt to pursue the implications of quantum mechanics beyond science and into all the neighbouring realms of our knowledge. He argues that what we have discovered about the fundamental nature of matter is so radically opposed to our everyday forms of knowledge that nothing can remain unaffected.

For Bohm the central change that must occur is that we must abandon old, fragmented ways of understanding the world. Classical science has conditioned our language to conform with its understanding of the world. This view, as I have shown, consists of a breaking down of problems into discrete, experimental and observable parts. We acquire our knowledge of the world by seeing Galileo's falling weights as a separate phenomenon from the air pressure which, in reality, gives us a 'false' reading. Equally, we see our own minds as irrevocably separate from the world they attempt to interpret.

The whole of quantum theory seems, to Bohm, to undermine this view. Non-locality indicates that particles can affect each other over immense distances. In some sense we do not yet understand they may actually be one system. To see them as different is merely a useful generalization. Yet to see them as the same appears to deny all that we think we know of the world. Bohm has argued that this is precisely what we must accept, whatever the difficulties: '. . . our notions of order are pervasive, for not only do they involve our thinking but also our senses, our feelings, our intuitions, our physical movement, our relationships with other people and with society as a whole and, indeed, every phase of our lives. It is thus difficult to 'step back' from our old notions of order sufficiently to be able seriously to consider new notions of order.'[11]

The logic behind such thought is clear: the world is made neither on the basis of any model derived from our common sense, nor from our classical science. It is fundamentally alien to both. Yet our lives and our culture are based precisely on these inaccurate models. There is, therefore, a discontinuity between the way the world 'really' is and the way we behave as if it were. And this discontinuity may be at the root of all our woes. The new science has an urgent message for us, here and now in our everyday lives.

Bohm is employing a central idea of intellectual and spiritual history: that we are in some way 'wrong'. Its most obvious contemporary form lies in the image of ourselves as polluters and wreckers of the planet. But, underlying this and all its other current expressions, is the conviction that there is something wrong with the way we 'know' things. This modern anxiety is an obvious descendant of the Christian notion of original sin. Both imply that there is an intrinsic flaw in our relationship with our world and both suggest that this flaw arises from a corrupted form of knowledge. The Christian corruption arises from our pride that made us wish to know more than God intended. The scientific corruption arises from the method of separation and dissection, of isolating parts of the whole in order to understand them. This has made us destroyers of nature. In order to understand the world we have to cut into it or break it open. We must create artificial conditions. What we do not do is passively contemplate the whole. We do not reverence creation, we break it open like a child to see how it works. And we cannot put it back together again. Classical science has not merely misled us in our laboratories and observatories, it has made us wrong in our world.

In the past – for example, in the minds of nineteenth-century romantics – such an argument would have been based on feelings, sensibility, emotions; but, in our age, the overthrow of mechanistic systems and their replacement with strange theories that defied common sense appear to provide something more – a form of hard evidence that the world was not the cold, simple place science would have had us believe. In essence, we could, perhaps, find value in the very place the Enlightenment had convinced us it could not be found: in the scientifically observable facts of this world.

As I outlined in the last chapter, this prospect had alarmed Max Planck, the founder of quantum theory. He saw the possibility that his own scientific insight might undermine science itself. In overthrowing previous theories, it potentially suggested both that all science was no more than a temporary culturally determined hypothesis. Worse still, the nature of quantum theory was such that it evoked a world of seething uncertainty and strangeness. The 'real' world on which the idea of science was based might be a fiction like any other. All of which could leave science as potentially vulnerable as religion.

'Is there any rock of truth left?'[12] he asked.

Planck's response was classical and tragic. Religion had been undermined by the assault of science, but, at least, we had been left with the consolation of hard truth. Now even that might be threatened. Werner Heisenberg took the more radical view that Enlightenment philosophy would simply have to be modified. He spoke of 'relativizing' the Kantian *a priori* and of dismissing the idea of the reality of things in themselves – an atom was a product of observation, not a thing in the world. The saving stability behind this, for Heisenberg, was that these new insights pointed towards a world constructed on a Platonic ideal of symmetry and balance rather than on a mechanistic ideal of cause and effect.

The anti-classicism of Heisenberg can be seen to lie behind David Bohm's desire to change all that we are on the basis of quantum mechanics. Because we are heirs of the Enlightenment and because the Enlightenment was founded upon science, then, since classical science was wrong, we are wrong.

Heisenberg can also be seen to inspire more populist and uncompromising speculations about the nature of our knowledge. Fritjof Capra, an American physicist, has produced the most explicit and popularly successful attempt to redefine science. He begins from the condition of original sin rooted in the fragmented, 'scientific' understanding:

The belief that all these fragments – in ourselves, in our environment and in our society – are really separate can be seen as the essential reason for the present series of social, ecological and cultural crises. It has alienated us from nature and from our fellow human beings. It has brought a grossly unjust distribution of natural resources creating economic and political disorder; an ever-rising wave of violence, both spontaneous and institutionalized, and an ugly, polluted environment in which life has often become physically and mentally unhealthy.[13]

Fragmentation is the human, moral consequence of the experimental ideal. In Capra it provides a rhetorical generalization of all our anxieties. New physics, however, tells us that fragmentation is not true of the real world. It speaks of harmony and interdependence. It tells us that, in a scientifically meaningful sense, everything is touching everything else.

Capra then draws elaborate parallels with the insights of eastern religions – in the unified visions of Hinduism, Taoism and Buddhism. These parallels are, for him, decisive. They show that the oriental sages long ago attained the truth towards which our corrupted, sinful science is only now beginning to lead us. We have paid a price for the material success of our classical science – we have been spiritually depraved. The new realizations can save us:

I believe that the world view implied by modern physics is inconsistent with our present society, which does not reflect the harmonious interrelatedness we observe in nature. To achieve such a state of dynamic balance, a radically different social and economic structure will be needed: a cultural revolution in the true sense of the word. The survival of our whole civilization may depend on whether we can bring about such a change. It will depend, ultimately, on our ability to adopt some of the yin attitudes of Eastern mysticism; to experience the wholeness of nature and the art of living with it in harmony.[14]

Like Bohm, but with fewer qualms and qualifications and with less precision, Capra directly connects physics with the human world. His holistic fervour sees no discontinuity between classical science and violence on the streets, nor between quantum mechanics and the wisdom of the Tao. He rejects not only the fragmenting tendencies of traditional science, but also the wisdom of Hume and Kant who saw the dangers and, indeed, impossibility of locating values, meanings and God in the world of the senses. In doing so he offers a kind of paradise in which our material system of knowledge leads us back to an ancient spiritual perfection, just as, in *2001*, the voyage to Jupiter leads us back to the condition of visionary foetus.

This is the opposite route to science-wonder to that of Feynman. Feynman dreamed of a truly scientific age when knowledge, traditionally understood, would be its own justification and our primary consolation. It need not provide any weird fireworks; it need only be knowledge.

But Capra wants that knowledge to go somewhere – specifically back to the wisdom of the oriental masters. Both views spring directly from the strange and complex new directions of twentieth-century science. But, whereas Feynman regards those developments as sufficient proof of the value of the whole scientific enterprise, Capra uses them in an attempt to convict that enterprise of an infinity of crimes and to bury it beneath a new spirituality.

The problem with Capra – and any number of other attempts to detect God in the quantum, relative and chaotic world – is that the spiritual language appears vague, drained and bloodless next to the scientific. Generalizing almost a century of quantum mathematics as displaying 'harmony and interrelatedness' has, perhaps, some meaning, tenuous though it may be. But using those words to connect that to the insights of oriental religions has none. In what sense can Lao-Tzu or the Buddha have 'known' what Heisenberg or Bohr 'knew'? In the sense, Capra would reply, that both these separate traditions had arrived at an underlying truth about the nature of the world. Their methods were utterly different, but the fact that there was a correspondence in the vocabulary of their conclusions surely amounts to corroboration.

Unfortunately, it is impossible to see how the vocabulary amounts to any more than wishful thinking. It is as if Heisenberg and Lao-Tzu had been discovered to agree that it would be best if we were all nice to each other and this was used as the basis of a hypothesis that the quantum and the Tao were one and the same. They may be, but we currently have neither the vocabulary nor the knowledge to make that correspondence meaningful.

And, in any case, what happens when physics changes again, as it surely will? Indeed, this is the most damning indictment of all these attempts to reconstruct values and meaning from within science. For science has not only become strange in this century, it has also become highly volatile. Values and meaning cannot be volatile, they must endure. If nothing in science endures, it can offer no meaning.

I apologize again for this breathless and intensive rush of speculation. But it was necessary to see that new ambitions have been born of science's new role in the world and its new radical theories. The variety of these ambitions and the excitement attached to them indicates, I think, two things. First that there was a need and, secondly, that people do, instinctively, know that the issue of science lies at the very heart of our culture.

But I can simplify the picture a little to try and establish what all this excitement means.

At one level it is a symptom of the struggle of our primary form of modern knowledge to cope with the almost unthinkable. Science discovered in the twentieth century that it was quite possible to be wrong in spite of being effective. Newton had been – and still is – spectacularly effective. But he was wrong. The same principle applies to a number of other areas of science. What had once seemed experimentally established beyond all doubt had been overturned What we had taken to be the Truth had turned out to be no more than a rough, workable generalization.

At the height of the nineteenth century's scientific and technological confidence, the answer to the question 'What is Truth?' would have been simple: the universal machine of science is truth. But after Planck, Einstein and Heisenberg, can anybody still say as much?

Your answer to that question is, finally, a matter of opinion. Many scientists and non-scientists alike have taken the view that the world mechanism is no longer a meaningful way of understanding reality. The vast, colourful carnival of speculation that I have attempted to describe in this chapter is the result of that kind of answer to the question. Equally, as I have said hard science can perfectly well continue to defend its essentially mechanistic and deterministic position by arguing that anything that seems, for the moment, to escape the net of science, traditionally conceived, is a function of our ignorance. Einstein stands as the exemplar of this view with his conviction that the indeterminacy and paradoxes of quantum mechanics would disappear once we knew a little more.

The point is, as always, not the answer but the fact that the question has been put. The transformation and acceleration of science in our century have created a new state of affairs. Science has found itself led into areas of thought usually associated with philosophy. Most importantly, it has found it increasingly difficult to retain the traditional, objective model of scientific inquiry. Developments such as quantum mechanics and the Anthropic Principle have challenged the central fiction of classicism – the idea that we could regard ourselves and the universe from outside, as gods of objectivity.

Such considerations have inevitably led people – like Bohm, Sheldrake and Capra – to speculate that we might have been too eager to take the bleakest of all views. The whole scientific effort which resulted in leaving us cold and alone might have been denying not only our instincts, but also the whole truth. We were wrong to see the universe as billiard balls and ourselves as accidents with no meaning or point. So perhaps in the quantum vibrations, the morphogenetic fields or the Tao we might find a reason to live.

Certainly the complexity and variety of speculation in our age suggests we have been liberated. The loss of mechanical certainty has freed us from a particular type of authority in precisely the same way that the loss of the intellectual power of religion and its institutions freed the human mind during the Enlightenment. So we have a playground of speculation about what reality might or might not be.

Yet, if this is liberating, it is also chilling. If everything can be called into question by theories as radical as Sheldrake's, does this mean that all our knowledge amounts to nothing? That all our science has not even scratched the surface of reality?

This in turn raises the question – as it has been raised many times before – of what our science actually is. In some ways our century has made this a particularly awkward issue for scientists There are, for example, two views about the present condition of physics: that it is in its most creative and powerful phase or that it is in a state of extreme decadence. The view that contemporary physics is decadent arises because so much theory is so far removed from any possibility of experimental verification. Superstrings cannot yet be shown to exist outside the equations which require them. Equally, the single unified force that supposedly existed when the universe was 10^{-25} seconds old is pure speculation and yet it forms the basis for much of the cosmological theory which lies at the heart of the most exciting and seductive tales of wonder emerging from science.

And, because of the remoteness of these theories from experiment or observation, mathematics has taken on a dangerous importance. The beauty and harmony of mathematical systems are being taken as evidence of the cosmologies to which they are attached. This leads physicists to cling irrationally tightly to their theories. Some point out that, when contradictory observations do arise, beloved theories are artificially modified rather than discarded.

To those who see decadence in all this, these over-mathematized physicists are reviving the Ptolemaic method. Ptolemy started out from the conviction that the earth was at the centre of the universe and then proceeded to devise an astronomical model around this certainty. The result was a fabulously complex system that worked if only because Ptolemy had gone to such extraordinary lengths to ensure that it did. The Ptolemaic system remains one of the great achievements of the human imagination. But it was never 'true' according to our standard of truth.

Theories of Everything, the great unification efforts that have sprung from the later work of Einstein, all share with Ptolemy's model an extravagant initial assumption – that there *is* a unity. There are two ways of saying this is absurd: first we may point out that reality may contain no such unity or, secondly even if it did, there is no guarantee that such a unity would be accessible to human reason, that it would be algorithmically compressible. Alternatively, of course, we may say that the attainment of unity may be perfectly possible, but it will be just another human dream like the rest of science; we will simply have created a reality that is conveniently unified.

Perhaps the answer to all this is that the very word 'science' has escaped its cage of meaning. Perhaps superstrings and other exotica belong to a realm of speculation more properly called metaphysics or even philosophy. Perhaps all this is the beginning of a process whereby a whole region of science detaches itself and becomes something else.

That may be so, but it should not make us overlook one underlying point. For traditional science, experimentally verifiable and technologically realizable, continues to shape us and our world. These wilder speculations are not yet of immediate importance. I hope to show, in my final chapter, that they are a symptom of something that is. But, for the moment, real science, ordinary science, devoid of philosophy and practised by people who have little time for these abstractions, continues to be the challenge and the threat.

8

The assault on the self

. . . to risk unreservedly being oneself, an individual human being, this specific individual human being alone before God, alone in this enormous exertion and this enormous accountability.

Kierkegaard[1]

We can say, of course, that there is nothing of real substance in all this speculation, these ultimate equations and these earnest attempts to spiritualize science. What does it matter what science says or does? We live on day by day. Exotic journeys to the beginning and end of time and space might be interesting, but only in the distant way that an alien religion might be interesting. As if we were anthropologists we might consider with wry detachment what these scientists are thinking. We might admire their wisdom, but always with the private conviction that it has no purpose for us. Science so far removed from practicality or point is an ornament, curious but not urgent. We are happy to stay at home and think about ourselves.

Unfortunately our forms of knowledge are not so easily contained. There is always a connection between the way we dream and think and the way we are. We are fooling ourselves if we think the state of science today is none of our concern. Just as the theology of the Roman Church was once of urgent significance to the people of Europe, so the condition of science matters to us. We are all connected by the problem of what we mean.

Science, like religion, provides a meaning that connects all it touches. But it is a more limited meaning that offers a form of truth without significance. And we cannot say that its abstractions are of no great importance, for it was the abstraction of quantum theory that gave us electronics, and of Maxwell's musing on the nature of Faraday's fields that gave us radio and television. If we think television has changed the world – which it has – then we must accept that science has done so a million times more radically for television is only one small aspect of what it has done. Equally, its abstractions signal the forms of belief within which we live as well as the real ambitions of our scientist-priests.

These ambitions have always faced two important barriers to the possibility of a complete scientific understanding. The first was the 'Why?' of the universe. I have already discussed the way science has begun to try to cross this frontier. The second barrier, to which I now turn, is the frontier of the human self.

We have seen how the scientific effort moved from the cosmos, through the living world and, finally, to man's inner nature, most potently and recognizably in the work of Freud. The pattern is almost absurdly neat: man is shown his place in the universe, then in the living world and, finally, in the contents of his own psyche. That, of course, is the classical view of the process. A post-classical version would be: *man shows himself* his place, he invents a place.

The last step, in either version, remains contentious: psycho-analysis may not truly be a science, it may be wrong. Nevertheless, its primary implication and ambition are clear: the human self will be drawn into science by the creation of a coherent causal chain of narrative to explain its existence and character. The self will become as clear as the Newtonian cosmos or the Darwinian story of organic variation.

The first and most important point to make here is that this has not happened yet. Neither at the level of the individual nor of societies has science succeeded in producing theories, predictions or analysis that have proved anything like as effective as those in the physical realm. None of the so-called human sciences – anthropology, psychiatry, psychology, sociology – have achieved the accuracy and explanatory power of the science of nature. As a form of imaginative literature they may occasionally attain a different kind of truth – notably in the works of Freud himself or in the sociology of Max Weber. But, as science, they have, at best, achieved a limited degree of statistical knowledge; at worst they have merely borrowed the authority of science to flatter simple opinions. There are no quantum mechanics of the psyche nor a relativity theory for society.

There are two conclusions that can be drawn from this. Either it has not happened yet, but it will one day when we know more; or it has not happened because it is intrinsically impossible.

There is a crucial difference between these two conclusions. To believe in the first is to believe in the possibility of ultimate human knowledge and in the essential correctness of the scientific quest. To believe in the second is to limit science.

Indeed, it might be said it is to cripple science. After all, it is reasonable to take the view that the self is all there is, and, therefore, to exclude science from that realm is to exclude it from everything. What is at stake is our ability to cross the inner frontier of the self. What is at stake, in other words, is not less than everything.

What we mean by self, however, is confused. To examine the matter is either to find oneself examining everything and, therefore, nothing, or it is to pursue some fantastically evasive fragment which vanishes every time we come close.

The basic Christian position is that we have bodies and souls. There have been different Christian definitions of the relationship between the two and of their respective qualities, but the essential point that there is an inner and outer reality is clear. In most conventional Christian eschatology, the relationship between the two is both temporal and moral. The soul is immortal; the body mortal. The soul is the infinite part of our beings; the body our finite connection to the changing, decaying world. The moral connection is that our souls are damaged or enhanced by our conduct in the finite world of the body. Though our soul is immortal, it is vulnerable to change – exhaltation or corruption – within the confines of mortality.

In medieval Catholicism the salvation of the individual soul was conceived as being uniquely within the gift of the institution of the Church. Part of the reason for the importance of Transubstantiation in the Counter-Reformation was that the sacrament of the Real Presence could be obtained exclusively through the priesthood. The completion and fulfilment of the human self was, therefore, only conceivable through its acceptance of the divine authority that was manifested in the Church. No other course was available.

The corruption latent in this absolutism seems all too obvious to us today. Total power was granted to a human organization. No earthly might could stand in its way, since the souls of any who rebelled were guaranteed to suffer in eternity. Fallible human beings might sin or even believe otherwise. Indeed, even within the Church disputes may occur. But the only complete answer to the issues of life was possessed by and in the gift of the priesthood.

The complex phenomenon of the Reformation can be explained or categorized in any number of ways – political, economic, intellectual or simply religious. But its central message was obvious: salvation could be attained outside the rigid systems of the Roman Church. In the history of the human self, this is a decisive assertion: it placed new significance and new responsibilities on the choosing, suffering, individual soul. The drama of salvation was internalized. This did not happen at once and nor was it a consistent phenomenon – Calvinist societies in particular were notable for their harsh, external control of individual lives – but the long journey of Protestant thought is a journey inwards. It culminates in the passionate anguish of Sören Kierkegaard's insistence on the primacy of the self choosing and acting forever beneath the gaze of God, a confrontation and a solitude unmediated by the pomposity of institutional faith or external rationality.

'Christian heroism,' he wrote, 'and indeed one perhaps sees little enough of that, is to risk unreservedly being oneself, an individual human being, this specific individual human being alone before God, alone in this enormous exertion and this enormous accountability.'

Even within Catholic orthodoxy the individual was beginning to flex his muscles in the face of authority. Artistically, the Renaissance was an assertion of the power of individual genius and, philosophically, it was a celebration of the rediscovery of classical humanism, of belief in the power of man as man rather than in his dependence on God and his impotence before nature.

I have already written of the way science sprang from and sanctified the authority of the individual insight. Protestantism and the Renaissance had effectively prepared the way: the first by insisting on the moral centrality of the individual and the second by its celebration of heroic humanism. But there was a price to pay for scientific individualism. The price was the expulsion of the self from the world. For science made exiles of us all. It took our souls out of our bodies.

This tendency is evident in the primary philosophers of the Enlightenment. Descartes provided a philosophical correlative of Protestant internalization by locating the only certainty, other than God, in the inwardly perceived thought processes of the human mind. Kant removed the real world beyond the possibility of ordinary human knowledge. Both placed the world that was the object of scientific investigation beyond the realm of the self. The key paradox of the modern was established: science was everything we could logically know of the world, but it could not include ourselves.

It was not simply that science was revealing the universe as progressively less benign; it was also the fact that it exposed no need for God and no dependence on man. The more we know, the less we appeared to have a role. The world worked without us, so the self, finding no way of defining itself in the world, retreated inwards – if the magnificent aggression of Kierkegaard can be called a retreat. What we became in ourselves was precisely and only that which science was not.

Meanwhile, on the other side of the equation, science executed its extraordinary metaphysical sleight of hand. It constructed a vision of the world *as if we were not here*. Hard, deterministic science's view of man is that he is a curious accident. Self-consciousness is a problem, but not of a different order from other problems and it is always misleading to believe that it is. The track of the scientific self is thus to step outside all other selves. The observing scientist observes as if he were a super-human, watching from outside the universe. He aspires to an additional reflexivity of consciousness, to a condition of self-self-awareness.

The defences against this astonishing and terrifying leap of the human imagination, apart from radical or conventional religion, demanded equally wild solutions. The self could reconstitute itself in terms of classical tragedy – lonely and self-determined in a friendless and heartless universe. Romanticism, of which Kierkegaard was a part and from which sprang Nietzsche and Freud, rediscovered the self in the revelations of its own drama. So much wonder had been stripped away from the objective world that it became necessary to elevate the subjective to the condition of art. In late romanticism – most familiarly with Oscar Wilde – the life became the work of art because all else had been conquered by the inartistic power of reason. And, finally, it is not a long step from the self-examinations of the romantic poets and painters to the narratives and myths of psychoanalysis.

The point was that, as scientific knowledge progressed to colonize the entire universe, the self became a safe refuge. In here we were safe from invasion and we found a home, an escape from the eternal wandering offered by science. It is vital to understand that this is no rarefied intellectual or aesthetic game that I am describing. The refuge of the self is something we employ as a defence at every passing moment. In our daily lives there is the ontological discontinuity felt by every individual: we feel different from the world and we derive both consolation and terror from this difference. We are consoled by the solitude of knowing we can never quite be explained, invaded or controlled by the world; we are terrified by the sense in which the world is indifferent to our fate.

We habitually assume there are two worlds: one that is ourselves and one that is not. Scientific knowledge exists for us only among the things that are not us. Newton or Einstein may tell you the world is thus and you may believe them. Or you may look at an anatomical picture of the brain and think that you have one of those inside your head. But it is never like that inside your mind; it is not even remotely like that. As Mrs Einstein said of relativity: 'It is not necessary for my happiness.' There is something else, something not captured by the equations and diagrams and this appears to be the feeling of being oneself.

This discontinuity between these two worlds may be crude, absurdly obvious, but it is overwhelmingly important for it emphasizes the way in which science can be viewed as telling us everything *except* what matters to us. Even if it tells us why we are here in a cosmological or biological sense, it does not do so in a moral sense. It does not even tell us why it feels like this.

And, finally, it has proved singularly incapable of telling us who we are. It has, in short, succeeded in telling us nothing about the self.

This is the deeper version of my point above about the general failure of the human sciences and it produces opposing responses roughly under the same headings of inadequacy and impossibility. The inadequacy argument would say that the various attempts to produce a science of the self simply have not been good enough.

Marx, for example, would have said that all our uncertainties, whether cosmic or ontological, arise from our place in the material cycle of history. Speculation is pointless since that too is rigidly determined by our economic condition. The only way forward is practical action to transform the world. This was based upon his supposedly scientific analysis of history. Freud, a pessimist, said the best we can hope for is an accommodation with all the conflicting stories of our biographies and our civilization. But any science in Marx was overwhelmed by the metaphysical faith in socially redeemed man and Freud's work represents an interpretation rather than a repeatable and transmittable system of self-knowledge. In neither system did prediction, one crucial touchstone of scientific validity, prove possible. Marx was wrong when he predicted that Britain would be the first nation to undergo a communist revolution and, indeed, wildly wrong about the development of economic history over the ensuing 150 years. Freud's generalizations, meanwhile, though poetically convincing can never be used to forecast psychological conditions in advance. They can only describe the past.

But such failures to turn the self into science need not be conclusive. The hard scientist can calmly reply that our science in the realm of the self has simply not been good enough. Perhaps, he might add, it is merely the complexity of the human realm which makes true science difficult. If we knew enough about an individual or a society we could predict with accuracy. The task is difficult but not intrinsically impossible. Freud and Marx, in this view, were just premature – they were trying to run before they could walk.

The second way of understanding the failure is to say that the wrong questions have been asked. All that Freud and Marx and their successors have been doing is attempting to extend the principles of classical science into the human soul. We are examining our selves as if they were falling weights or curved starlight. The error can seem bewilderingly elementary. A moment's thought would convince a child that the most striking thing about us that that we are utterly unlike anything else in nature. Light, gravity, even the whole biological realm are related to us only in the most superficial way: we reflect light, if dropped we fall and we have a bodily system roughly comparable to a large number of animals. All of which is trivial compared with the one attribute we have that is denied to the rest of nature – self-consciousness.

In a sense, this makes the curious failures of science more understandable. For the problem of the self is not, initially at least, its origins, its history or even its mechanisms. The problem is its existence in one species out of millions and, apparently, on one planet out of billions. We know and we know that we know; animals only know and cannot, therefore, be said in any meaningful sense to have a self. As far as our cosmic status is concerned, all that we can say for the moment is that we appear to be alone. Even if we find that we are not, it will remain clear there is something very rare and odd about the fact of consciousness.

Hard science will fight back at this point by attempting to deny this is a problem at all. Self-consciousness is merely a by-product of evolutionary complexity. Animals develop larger brains as survival mechanisms. Over millions of years these brains attain awesome levels of miniaturization and organization; indeed, they become the most complicated things in the universe. Then, one day, this complexity gives rise to something utterly unprecedented. Perhaps the internal functional explanation is that the brain-machine becomes so complex that it begins to make new connections not directly related to the daily requirements of survival. By some design fluke, a surplus of processing capacity emerges which manifests itself as self-awareness. The higher primates are able to start the thought processes which will lead to the cosmically staggering insight: 'I am a higher primate.' Perhaps the critical moment comes when, as the biologist Richard Dawkins has suggested, the brain achieves sufficient complexity to be able to contain a model of itself. Or, as Douglas Hofstadter puts it, 'The self comes into being at ᵗhe moment it has the power to reflect itself.'[2]

Or perhaps, externally, the explanation is the slow evolution of a quite new tool: language. Primate grunts become more specific and more versatile until words emerge. The trick is so successful that words generate more words and these, in turn, generate new meanings which feed back into the brain and produce thought.

Such are the perfectly respectable speculations of hard science as to how consciousness might arise. As a fact in the world it may not have the obvious evolutionary role of legs or wings, but it can be understood as a kind of feverish indulgence in excess on the part of evolution – there are, after all, other examples of non-functional, indeed positively unhelpful animal attributes that have occurred because of the blind statistics of deep time. The self is like the peacock's tail.

The reason such explanations feel inadequate, even though, as children of the scientific age, we probably accept them at the back of our minds, is that they are incoherent. They do not explain self-consciousness, they explain complexity. What is unprecedented about the human mind is not its processing capacity, but the existence of, for example, the colour green. We can understand how a brain might evolve and work and what it might be like. But that will tell us nothing about the existence, apparently in such a brain, of a concept like 'green'. What is it? What is it there for? What does it do? What does it feel like? What does it mean?

Evolutionary explanations of consciousness may or may not be correct. But they will remain incoherent unless precisely what is being explained is defined more closely. And this, of course, runs into the problem that what you are trying to define is what you are. It is by no means clear that this can ever be possible. Indeed it may be logically impossible. Any such definition would have to be in language and yet it would also have to be *of* language. It would be like asking a photographic film to take a picture of itself. The problem with language and self-consciousness asked to perform such a task is that they produce nothing more than a series of infinite regresses as if parallel mirrors were built into our minds. The only certainty may simply appear to be the

Cartesian one that self-consciousness includes the capacity to wonder what self-consciousness might be.

Of course, the hard evolutionist may still respond by claiming that this is a by-product of complexity. The elaborations and anomalies of our language and our awareness are merely a kind of surplus capacity to idle that happens to occur in the brain. We have more neurones than are strictly necessary to gather food or reproduce so, when they are not thus engaged, or even sometimes when they are, they chatter on in endless circular arguments which only *seem* important. In reality, they are trivial – in the words of Peter Atkins they are 'special but not significant'.

But, again, this is incoherent. How can it be 'not significant' that we are able to use and understand the words 'not significant'? What meaning can the word 'significant' have in such a context. Significant to what? If self-consciousness is 'not significant', then where on earth is significance to be found? Atkins is trying to attach shock value to the phrase 'not significant'; but the effort is in vain, for we know what it means as well as he does and it cannot be made to mean this.

Atkins' problem, *our* problem, is that scientific knowledge is fundamentally paradoxical. The paradox is that all of science's 'truths' about the 'real' world are based upon the most flagrant distortion. In creating an understandable universe, we have committed ourselves to the most gross and obvious oversimplification. We have excluded the understanding mechanism, the self.

The point was summarized by the quantum physicist Erwin Schrödinger. 'Without being aware of it,' he wrote, 'and without being rigorously systematic about it, we exclude the Subject of Cognizance from the domain of nature that we endeavor to understand. We step up with our own person back into the part of an onlooker who does not belong to the world, which by this very procedure becomes an objective world.'[3]

Hannah Arendt made the same point when she wrote that science acquires its knowledge 'by electing a point of reference outside the earth'.[4] And the physiologist Lord Adrian pointed out that the very act of contemplating ourselves obliged us to step outside 'the boundaries of natural science'.[5]

Schrödinger also quotes Carl Jung: 'All science, however, is a function of the soul, in which all knowledge is noted. The soul is the greatest of all cosmic miracles, it is the *conditio sine qua non* of the world as an object. It is exceedingly astonishing that the Western world (apart from very rare exceptions) seems to have so little appreciation of this being so. The flood of external objects of cognizance has made the subject of all cognizance withdraw to the background, often to apparent non-existence.'[6]

I include all these different sources making the same point to indicate the enduring and crucial nature of this puzzle. It is not some remote, analytical game of words; it lies at the heart of what we know. For all these thinkers were defining the way in which science's faith in the objective investigation of a real world is a metaphysic like any other. What was found to work by Galileo and his successors was a way of viewing the world *from outside*. In discovering the surface of the moon was not what the preceding, Thomist, rationality would have required, Galileo was, we might say, being objective. But this is a confusion of perspective. What we should say is that Galileo at that instant *invented* objectivity. He invented a method which said that, in effect, the moon is the way it is with or without human connivance or relevance. This may seem obvious to us now. But, again, try the thought-experiment of seeing the night sky with innocent eyes. Suddenly it becomes probable, indeed obvious, that it does mean something to you, that you are implicated in this majestic vision. But, for the purposes of science, you must deny that understanding, you must see the cosmos out there and beyond. Galileo's discovery – or his invention – was that an extraordinarily effective way of understanding the world is to pretend that we do not exist.

Few faiths, cults or institutions can have made such a bizarre and extreme demand of their adherents. It is precisely as if some sect had insisted only that its followers believe they were invisible and all else would follow. Such a faith would be confined, we assume, to a few eccentrics and inadequates. Yet science's demand is even more extreme and we do not notice our own acquiescence, our own eccentricity. And we do not notice because, astonishingly, the demand produces results. It works.

There is a further reason why we do not notice. Descartes invented more than a method for doing science, he also invented a conception of the human body as a part of the objective world.

We can examine our bodies much as we would any part of nature. They are, we assume, constructed on the same principles and they are, therefore, mechanisms. The Cartesian soul occupied this body rather as a pilot does an aircraft. This is convincing because our bodies do appear to be like the rest of the nature: heated they burn, dropped they fall. So science does not demand that we remove all of ourselves from the world, merely our self-consciousness. Our bodies dwell in the realm of the objective, our minds/souls elsewhere. Except, of course, there can be no 'elsewhere' in science. Yet, in spite of that, we allow our souls to be extracted from our bodies.

Descartes avoided the catastrophic implications of this view by employing a benign God as a bridge back to the world, as a guarantee that we were not quite as paralysingly alone, ignorant and impotent as his dualism might have suggested. The weight of scientific progress made the bridge collapse, leaving behind a paradox – the invented objectivity of the world. But, in the absence of Descartes's God, the human body becomes a special sort of problem. Here it is, this mechanism, partaking of the laws of nature and yet 'I' – wherever I may be – control it.

Schrödinger wrote in a sentence whose contortions seem to mirror the knots of thought which he was trying to undo: 'I in the widest meaning of the word, that is to say, every conscious mind that has ever said or felt "I" – am the person, if any, who controls the "motion of the atoms" according to the Laws of Nature.'[7]

In this context, the appeal of certain aspects of quantum mechanics becomes obvious. Quantum theory suggests that there is, in fact, some profound interaction between the observer and the world. Objectivity appears to vanish at the sub-atomic level. We only see what the entire structure of apparatus, observer and observed determines we can see. The observed is not some irreducible absolute; it appears, rather, to have little definite reality other than in the terms of the observer. The quantum appears to be telling us that we are *in* the world whether we like it or not, that objectivity is an illusion, a fiction that only works at the level of gross generalization. If this is true, then in the fine structure of the universe we may find ourselves.

Such a view, of course, is as fraught with danger as any of the religious conclusions derived from quantum mechanics that I discussed in the last chapter. We cannot rely on any science, including quantum mechanics, to continue to conform to our desires for a reintegrated sense of self. It is unlikely to stand still. New developments may occur that will restore the quantum, as Einstein expected, safe and sound to the bosom of classicism.

In any case, how meaningful is it to derive a new harmony, a new form of personal salvation, from the distant reaches of mathematics and experimental physics? What priests, what sorcerers would be required to bring back the message we longed to hear? The message is not transmittable. When Paul Dirac was asked what he meant by 'beauty' in mathematics or physics, he replied that if the questioner was mathematician then he need not be told, if not then nothing he could say would convince him. What kind of truth about ourselves would it be that could only be understood by such incommunicative specialists?

But what quantum mechanics, among other developments, has achieved is to focus the eyes of our age on the paradoxical role of the self in our forms of knowledge. The self, in our day, has emerged battered but finally, if ambiguously, victorious after a series of crushing defeats. Removed from the world, it had formerly survived only in tiny enclaves of theology and philosophy. In the nineteenth century it was nurtured and defiantly glorified by impatient geniuses like Kierkegaard and Neitzsche. The very fact that science has insisted on its expulsion from its version of nature made it the one safe haven – what was nothing to the mechanists was everything to the romantics.

Now this glorification of the self has become almost the primary defining quality of technologically advanced and affluent societies. As a contemporary safe haven, the realm of the self has become a kind of recreation, a holiday resort. A peculiarly intense variety of narcissism is the essential philosophy of mass communications. The forms this takes are instantly familiar: physical self-cultivation is pursued through dieting, exercise and exotic, often consciously anti-scientific, remedies; improved mental selfhood is pursued through personal advice columns, interminable analyses of 'relationships', a whole range of variations of psychoanalysis and, occasionally, fringe religions.

Much of this may seem remote from either the exaltations of

Nietzsche or the great ponderings of Jung and Schrödinger. But the connection is clear: the self has successively been discovered as a religious, philosophical and, finally, popular narcissistic refuge. In this final form the confusions reflect this difficult heritage. There is, for example, the ambiguity of science in the narcissistic imagination. Some of the paraphernalia of self-cultivation borrows the authority of science. Cosmetic advertising is always employing vaguely scientific language to assert the efficacy of the products and the whole popular landscape of mental health is dotted with psychiatrists and psychologists with their characteristic, if shaky, generalizations that tend to begin with the ominous and arrogant phrase: 'We have found that . . .'

In contrast, there is much anti-scientific sentiment. Science is frequently believed to be a crude and brutal destroyer. Alternative health therapies are constantly being sought, most insisting on the idea of the treatment of 'the whole person', an obvious popular surfacing of the old intellectual unease with the experimental basis of conventional science.

'High in calcium,' reads the text of a hot-drink advertisement, 'low in fat, free from artificial additives.'

The complex balancing act that preserves science as both destroyer and protector is revealed. Science tells us that we want calcium and we do not want fat. Equally, however, science has made things called 'artificial additives' which we also do not want. So science is simultaneously rejected and embraced. And these few words tell us even more. For the drink is, of course, utterly artificial – it has to be to ensure that it has much calcium and little fat. Yet we wish to reject the word 'artificial' because it implies a barrier between ourselves and nature of which we do not wish to be reminded. Science first constructs the barrier and then warns against its baleful consequences.

In this context, the word 'nature' takes on a crucial ambiguity. Science as an optimistic, acceptable pursuit will be credited with the expansion of our knowledge of nature. The word is here synonymous with the word 'world'. And the idea is that it is clearly a virtuous activity, science is, as it were, approaching nature in humility. But, in the modern context, nature tends to be opposed to the idea of science. Nature is understood as the world left alone, not interfered with. Science approaches nature with the arrogant ambition to control her, science is artificial as opposed to natural. In this version science loses its virtue. It becomes mistrusted, as it was once before, as allied to the black arts – the most 'unnatural' things that human being did.

This relates to a profound and historically important moral confusion. Clearly this is, to some extent, generated by commercial interests that wish to exploit our ambiguous sense of virtue. The attempt to persuade people that they *want* to do something is reinforced by the implication that they *ought* to do it. So, for example, the exaltation of the appearance of youth is the mainspring of almost all cosmetic and physical health marketing. People are persuaded they wish to look young for any number of selfish reasons – sexual, social or whatever.

The one-dimensional ignobility of such an aspiration, however, is not easily overlooked. So various strategies are devised for creating a moral dimension. Most crudely there is 'you owe it to yourself' as if morality could be reintroduced through some neurotic division of the personality into debtor and creditor.

More sophisticated and more insidious is the general impression that the pursuit of youth and health is intrinsically a good thing. At this point the simple commercial device can be seen in a somewhat grander perspective. First, the elision of physical well-being and morality clearly responds to the need for some kind of rationality. The self, deprived of a cohesive religious meaning and lost in a complex and vast social and technological system, retreats into self-cultivation as one way of attaining an identity – the one offered in the imagery and narratives of mass communication. But this cannot be rationalized other than as a defence mechanism. Narcissism can only be endowed with rationality by some form of conviction that it is the right way to behave. This rightness is absurd, of course, since there is no meaningful external standard being applied. But it is there as a rhetorical device. It becomes pernicious when it is extended logically to include the view that, if health is right, then sickness is wrong. This starts with activities that are known to be bad for you – modern taboos – smoking and over-eating, for example. A fat smoker is, in the cosmetic imagination, a bad man. This, then, too easily extends imaginatively to include any illness. This may never be acknowledged, but it is the inevitable outcome of the belief that health is good in any sense other than fortunate.

This elevation of narcissistic self-cultivation to virtuousness closes a circle. The self is denied is place in the world and its source of values. It resorts, finally, to a pagan act devoted to its own cultivation and worship. This, in justification, is said to be a virtue. The circle is closed. A morality is deduced from the facts of materiality. What the Enlightenment philosophers could not honestly do, we have chosen to do dishonestly.

The depravity of the process arises from the fact that it is only achieved by identifying the self with the relative condition of the body or, in the case of mental well-being, with its externalization in jargon and verbal hypnosis. This latter case involves the whole vocabulary of 'feelings', 'relationships', and self-analysis – journalistically known as 'psychobabble' – which has become the lingua franca of the developed world.

This might be seen as the final, desperate attempt to turn the self into a viable refuge. This vocabulary is a fevered melange of psychoanalytical terms and its justification derives from the moral inversion described above – it is *good* to cultivate yourself, in this case by talking. The intention appears to be a constant commentary on emotional states aimed at arriving at some equilibrium. By talking, it is assumed, what would otherwise remain private is exposed to a public, albeit intimate, realm where it can be dealt with. The reality is, of course, that the talking and requisite vocabulary create most of the dramas they pretend to describe. But the idea is that these words describe conditions that exist within the speaker's self.

On one level this phenomenon is an expression of the success of the idea of psychoanalysis as an objective, descriptive enterprise. At another level it seems to be a kind of neurotic avoidance of the idea of mortality in that the externalized self – locked for ever in a continuous process of analysed becoming – can never die. But its real significance is the same as that of the anti-smoking, anti-fat health fanatic. It lies in its notion of the self as a valid, virtuous haven, as the only realm in which our experience of the world can be said to have content or meaning.

But, as I said earlier in this chapter, the issue is whether the frontier that surrounds this haven can be crossed by science. Popular forms of psychoanalysis as well as health obsessions are, necessarily noncommittal on this point. They both believe and disbelieve in the ability of science either to invade or assist our inner natures. They require the best of both worlds so they sustain a balance of probabilities, a balance that is repeatedly demonstrated by the subtle ambiguities of advertising and marketing.

Within science itself, however, the efforts to cross that frontier have been growing increasingly determined. Mechanical, deterministic science could, of course, have continued for ever without making the attempt. Its functional position outside the universe was too effective to abandon. Freud's assault on the self was performed in the name of classicism, but could easily be seen, by classicists, as wildly premature, a risky, potentially discreditable undertaking,. What was needed before we attained his narratives was a more effective and convincing underlying map of consciousness. Twentieth-century science has not yet managed to make it clear whether any such map was possible or what it would look like. We may derive from quantum theory the view that any such detailed knowledge of the human self was destined to be indeterminate or, equally, we may derive from the astonishing accuracy of the same body of knowledge the view that it was perfectly possible.

There were spectacular developments like the unravelling of DNA that seemed to indicate the possibility of a mechanically comprehensible picture of the self. This was one twentieth-century achievement that seemed to restore mechanistic determinism safe and sound to scientific centre stage: in the DNA molecule we were told that we were able to see the blueprint for all that we were. It was out there, real, in the world.

But the contemporary issue of the inviolability or otherwise of the self is not being fought out under the microscope or in particle accelerators. It cannot yet be seen as a double helix or a quark. Instead we are searching for ourselves in numbers.

I have written in other chapters about the essential puzzle of numbers – whether they are 'real' in the sense that they reflect a true state of affairs in nature or whether they are merely a human construct, a convenience. Both arguments have an intuitive appeal. We feel numbers are real because it seems obvious that everything can be counted. On the other hand, it appears obvious that they are a human invention. For, if they are to be seen as 'real', then they must have, in some way, pre-existed us – a bizarre notion that implies that, when dinosaurs roamed the earth, eight multiplied by seven still equalled fifty-six.

Pythagoras was convinced of the transcendent reality of number, a faith that was badly shaken by the discovery of irrational numbers – the square root of two, for example. Nevertheless, this platonic faith was reborn with *Scienza Nuova*. The idea of arithmetic as an implacable absolute made perfect sense in a mechanical universe.

In the nineteenth century, however, ominous problems began to emerge in mathematics, just as they did in physics. The classical geometry of Euclid, for example, had long been assumed to be conclusive – until the discovery of the possibility of non-Euclidean geometries. Equally, set theory, notably in the hands of Georg Cantor, began to reveal strange paradoxes. The problem here was the essentially classicist assumption that logic and mathematics could be unified and this would represent a crucial unification of the philosophical and the scientific. Logic and mathematics should, ultimately, be the same thing.

A set in mathematics is precisely what it is in ordinary language – a group of things with something in common, as in the set of all things that are red. In the maths of Cantor and, later, in the logic of Frege, they seemed to offer powerful ways of defining number, of grasping the nature of mathematics.

But problems emerged, the most celebrated being the set theory paradox defined by Bertrand Russell. By applying the idea of a set to numbers, we can say that the number three, for example, is the set of all sets that have this property of threeness. The importance of this aspect of the theory is that it appears to provide a solid description of what a number is. Russell's paradox starts from the position that it is clearly possible in set theory to have a set which is a member of itself: sets of all things that are not red, for example. Such a set would not be red and would, therefore, belong to itself. But Russell pointed out the possibility of a set, which he called R, which is the set of all sets which are not members of themselves. We now have to ask if R is a member of itself. Clearly, if it is not, then it belongs to R. So R appears both to belong to R and not to belong to R. The reassuring solidity and completion of set theory – and, therefore, of mathematics as a whole – had evaporated as had so many other nineteenth-century convictions under the corrosive gaze of the twentieth.

Russell, along with A. N. Whitehead, attempted to restore solidity with *Principia Mathematica* (1910), a monumental work which posited a theory of types to replace that of sets. But taking these technical details further is not necessary. For Russell's paradox exposes the heart of the mathematical problem and it was not one that was ultimately to be solved by the theory of types.

The problem is the apparently inbuilt paradox of self-reference. This is much more easily expressed by the simple puzzle: 'The following sentence is false. The preceding sentence is true.' Pursuing the meaning of these words leads us into a closed loop from which there is no escape. Language, like mathematics, seems to contain these strange islands of circularity which cannot be made to conform to any neat, formal summary of the workings of either words or numbers.

But the intuitive sense that mathematics must be capable of completion and certainty is not easily denied. After *Principia*, mathematicians persisted with the quest to find the proof that their discipline could be defined as consistent and complete. It was a challenge explicitly set at the beginning of the century by the German mathematician David Hilbert and, for thirty-one years, it was the primary goal of mathematical theory. Then Kurt Gödel proved conclusively that the quest was futile.

The point about Gödel was that he showed that the problem of closed loops was not a passing curiosity, but a fundamental reality that would always prevent any arithmetical system from being complete. Understanding Gödel's proof is a hit-and-miss affair for non-mathematicians; sometimes it makes sense, sometimes it does not. But all that is really necessary to grasp its implications is this simple statement of its conclusion: 'All consistent axiomatic formulations of number theory include undecidable propositions.'

This is one of the most profoundly shocking or exhilarating

statements of twentieth-century thought. What it means is that we can find statements within any system which we *know* to be true, but we cannot *prove* them to be true. Thus there is a distinction between true and formally provable. Our knowledge is different from – more than – mathematical proof.

For those who had always intuitively understood mathematics as self-evidently complete and consistent, this was a devastating realization. Hilbert's question had been answered with a bewildering negative. In the outermost reaches of number theory we had discovered a door that was indisputably closed. It was an apparently final and incontrovertible limit to mathematics.

Or was it? And what has this to do with the problem of the self?

The connection is computers. Charles Babbage in the nineteenth century had designed mechanical calculating machines of extraordinary beauty and complexity. But it was not until the 1930s and 40s that electronics began to make a genuine artificial 'brain' seem possible. Accompanying this process, theories of computation were being developed, most powerfully by the English mathematician Alan Turing. In 1950, for example, Turing published a still highly relevant test as to whether a machine could be said to think. In this a computer and a human are hidden from view and an interrogator has to decide which is the human and which the computer simply by asking questions. If he cannot decide, then the computer must be credited with powers of thought.

But, more importantly, Turing produced a correlative of Gödel's theorem for computation theory. Also in response to the challenge of Hilbert, he worked on the question of whether there was a mechanical procedure for answering all mathematical problems. He conceived of the Turing machine, a simple computer, which would always stop when a solution was reached. He discovered that there was no way of knowing in advance whether the machine would actually stop.

Again the mathematics are not necessary, only a grasp of the implications. Turing had not shown merely that some mathematical problems were mechanically insoluble. He had show that whole groups of problems were intrinsically insoluble because we could never know in advance which procedure would lead to a solution – we could not predict whether the machine would stop. Again it is a division between true and formally provable. Most importantly it is a case of saying that the point is not whether we can construct an algorithm – a mechanical procedure – to solve a specific problem, but rather how we choose the right algorithm. It is as if in order to do maths at all, we have to step outside its procedures. If we cannot do this we literally cannot start to do mathematics. This appears to suggest that we have something a machine can never have – an ability to grasp truth intuitively in the absence of formal evidence. So our ability to 'see' mathematical truth is as relevant to the procedure as our ability to prove it. And this ability, in some way, precedes maths – it must precede maths in order for us to start doing maths. Maths cannot, therefore, be everything.

Any mathematical operation that a Turing machine cannot perform in a finite time is known as non-computable. Now, if the human mind is a computer, such non-computable operations are irrelevant. We can still potentially make a computer that thinks and is self-aware like us. But, if such functions do exist in the human mind, then we may never be able to build a machine version of ourselves. John Barrow explains the significance of this for machine intelligence: 'If . . . the action of the human mind involves non-computable operations, then the quest for artificial intelligence cannot succeed in producing computer hardware able to mimic the complexity of human consciousness.'[8] This does not, for Barrow, mean that maths is itself limited, nor that the universe is not written in numbers.

'If the universe *is* mathematical in some deep sense,' he explains, 'then the mysterious undecidabilities demonstrated by Gödel and Turing are part of the fabric of the universe rather than merely products of our minds. They show that even a mathematical universe is more than axioms, more than computation, more than logic – and more than mathematicians can know.'[9]

Apparently science as mathematics had proved that we were not the simple mechanical accident we had once thought. Rather we appeared to possess some ability to see into nature through mathematics that would always be denied to machines. Our ability to 'see' solutions rather than the need for proof was a mystery beyond the reach of any conceivable mathematical operation. Perhaps the self really is the one frontier that science cannot cross.

Unfortunately we cannot necessarily draw such a conclusion. To understand the next stage of this issue, we now have to enter the strange and confusing landscape of AI – artificial intelligence.

AI appears in a number of guises. At the practical, commercial level it is simply the name given to new computing systems which will make computers more flexible, usable and autonomous. Such developments are often intriguing, but, for the moment, they only appear to offer us an intensification of the current role of computers in our lives. More important for the issues of this book is the higher debate about AI which involves, in essence, the old sci-fi question: can we build a machine that thinks?

The fact of the computer has already convinced many of us that we can. These machines are particularly potent, modern symbols for the power of science. From the first mechanical calculators it has been quite common to speak of them as 'thinking' or 'working things out'. Now, with vast processing and memory capacity available to all, it is often linguistically easier to endow computers with semi-human attributes than to continue to refer to them as machines. In part this is because of the software-hardware parallel with our own images of ourselves that I discussed in the last chapter and, in part, it is because of the type of complexity involved – computers are not doing more than we are simply in the sense that a car or aircraft might; they are doing more, apparently, in our special field of calculation and organization.

They are like us and they have the additional similarity of inwardness – their work is invisible, we cannot watch gears turning or valves opening. Like us they do not disclose their operations.

Now it may be assumed, on the basis of this imaginative identification with their workings, that the gap between computers and our brains is simply one of storage and processing capacity. Computers have developed to their present level of sophistication by technological improvements that have led to bigger memories and more rapid processing. Such improvements will continue and, at some point, computers will attain the take-off point of self-consciousness. Clearly this possibility offers a kind of ideal test for the opposing views of the self as either a superfluity of complexity or an irreducible anomaly. If a self-conscious computer can be built, the first argument wins, if not then the second wins – we are more than machines.

The Gödel-Turing problem appears to support the view that no machine can be built that thinks. This is because the building of the machine is a mechanical procedure that involves choosing particular algorithms. Turing seems to show that the very choice cannot be made and Gödel that there exists a strange realm of the true but unprovable which would make it impossible mechanically to realize the full human quality of thought.

The development of chess-playing computers demonstrates an aspect of the problem. Early attempts were held back by the assumption that the computer had to be able to work out every possible future variation from each position. In practice this required an impossibly large memory and centuries of processing time. From the initial position the problem was most acute as, evidently, all possible combinations of the game were potentially present. The computer could not algorithmically compress the game, so it would tend to play every possible game before it made a move. Gradually the problem was overcome by building in horizons of possibility beyond which the computer did not calculate as well as ways in which it could analyse situations *in general* rather than in grinding detail. Now computers are as good as the best human players. Chess, however, is one thing . . .

. . . thought is another. The argument about 'Hard AI' – the most thoroughgoing AI view that says our intelligence can be mechanically duplicated – has a long way to go. Two huge books in recent years have come down on either side. The first is *Gödel, Escher, Bach: an eternal golden braid* by Douglas Hofstadter and the second *The Emperor's New Mind* by Roger Penrose.

Perhaps the most startling fact about both books – apart from their size – is the way they both feel obliged to include everything in the debate. Philosophy, physics, music, quantum theory, relativity and many other disparate fields are included to illuminate what begins as an entirely technical mathematical debate. This is not just a passing critical observation. The scope and ambitions of the books demonstrate that both Penrose and Hofstadter have realized what is at stake. In some final sense the inviolability or otherwise of the human self is the complete and most pressing issue of contemporary knowledge.

Hofstadter's vast and self-consciously eccentric approach supports Hard AI. His argument is partly concealed by technical detail and the wild extravagance of his method of illustration. But it is quite simple.

All the anomalies of mathematics and language are real enough and they all revolve around the problem of how anything actually starts. We cannot choose in advance the right algorithm for the Turing machine. For Hofstadter this is a problem of meaning. Our brain receives messages which it interprets. But, in order to do this, it must have a further level of a message which tells you how to understand the message. This suggests a complex hierarchy: at the top is the process of deciphering the contents of the message, below that the ability to recognize that it is a message, below that the ability to organize and differentiate our perceptions, below that the mechanism that links our perceptions to our processing capacity and, somewhere at the end of this chain, the individual electrical incidents in the fabric of the brain. But this hierarchy, in itself, explains nothing for it suggests an infinite regression. Once we are at the level of individual electrons, we can still keep asking 'Why?' How can we ever get started?

Here is Hofstadter's answer: 'This happens because our intelligence is not disembodied, but is instantiated in physical objects: our brains. Their structure is due to the long process of evolution, and their operations are governed by the laws of physics. Since they are physical entities, *our brains run without being told how to run*. So it is at the level where thoughts are produced by physical law that Carroll's rule-paradox breaks down; and likewise, it is at the level where a brain interprets incoming data as a message that the message-paradox breaks down. It seems that brains come equipped with 'hardware' for recognizing that certain things are messages, and for decoding those messages. This minimal inborn ability to extract inner meanings is what allows the highly recursive, snowballing process of language acquisition to take place. The inborn hardware is like a jukebox: it supplies the additional information which turns mere triggers into complete messages.'[10]

Hofstadter's is, therefore, a sophisticated version of the super-fluity view of self-consciousness – we have a mechanism which trips over into self-awareness at some level of complexity. But there is nothing unique about that mechanism, it is a physical system subject to physical laws. All the infinities, all the paradoxes end at this point. The brain is self-starting. There is no mystery, only a fabulous complexity which, in time, we shall master. In this context the experience of the colour green is no problem. We can look at it from inside the system – from the 'vortex of self-perception' of the brain – and we have the experience of greenness. Or we can look from outside the system and we may describe the precise wavelength of green light. The problem of the self is thus restored to classicism by saying that, of course, the problem is peculiarly difficult as the brain is quite obviously the most complicated thing in the universe. But the problem is not fundamentally different: 'We should remember that physical law is what makes it all happen – way, way down in neural nooks and crannies which are too remote for us to reach with our high-level introspective probes.'[11]

Penrose, paradoxically, is both more sober and more romantic than Hofstadter. For him the Hard AI position is riven with contradictions. He points out that the Hard AI belief in the algorithm represents a curious intellectual irony. For, in insisting on the importance of a combination of the abstraction of the algorithm and the materiality of the brain, they have revived a form of dualism with the algorithm replacing the soul. And dualism was precisely what the materialists, the forerunners of the Hard AI supporters were most concerned to eliminate. The last thing they needed was a soul.

But the key point is that, for Penrose, the unprovable-but-true state of affairs is decisive. When we 'see' rather than prove something to be true, that action cannot be coded into any formal mathematical system. Yet it is a moment in which we glimpse the underlying truth of the world. For Penrose is a platonist, convinced of the reality of mathematics and of our unique place in the universe as decipherers of its mysteries.

He defends our instinct that there must be 'something missing' from the mechanical-deterministic picture of the universe. Yet he runs into a problem here. Clearly we assume the 'something missing' is some kind of spiritual dimension and so it may be for Penrose. But, more precisely, he is convinced that there is something missing in our science which is holding us back. For Penrose believes our physics remains radically incomplete, specifically he thinks we require a theory of quantum gravity which may, one day, make it possible for us to 'elucidate the phenomenon of consciousness'. The paradox here is that Penrose may not actually be arguing against the Hard AI view, he may simply be saying that we do not yet know enough to pull it off . . . or he may not. In conversation with him I have found that this issue is certainly not settled in his own mind. The main point, perhaps, is simply that the current Hard AI arguments do not, for him, stand up. 'Computability is not at all the same thing as being mathematically precise. There is as much mystery and beauty as one might wish in the precise Platonic mathematical world, and most of this mystery resides with concepts that lie outside the comparatively limited part of it where algorithms and computation reside.'[12]

For the moment which side you take on the issue of Hard AI is an article of faith as it will be some years before a computer that can remotely be said to 'think' in any recognizable sense will be built. The Hard AI view is that it will, of course, happen and it is only our sentimentality that conceals from us the fact that something has happened already. To the hardest of Hard AI enthusiasts a simple thermostat thinks – in expanding or contracting the small piece of metal that turns our central heating on or off is performing a primitive version of what we do when we decide to do something and then do it.

Computers, in this view, are well down the road to full self-consciousness. Indeed, ultimately they will take over. Silicon-based evolution will succeed carbon-based evolution. We will download our personalities on to machines and become immortal by making any number of back-up copies of ourselves. Computers are, in fact, the next phase in the history of the universe. Consciousness will be freed from its fragile biological support system and be able to roam the universe without restraint. Science having conceptually removed our souls from our bodies will finally do so physically as well.

This is a modern incarnation of Hegelianism or liberal theology. Science is part of the unfolding narrative of the world and our transfer to a silicon mode of existence is simply the next phase. Liberalism consoles itself with the idea that this must be what is intended to happen.

If the Hard AI people are right, this development is not spectacularly far in the future. Computer processing power is growing extraordinarily rapidly. 'Human equivalence' at the present rate of growth could be attainable in a supercomputer by 2010 and in a personal computer by 2030. Some recent reports suggest this may happen much sooner. So the issue of a silicon self is with us now in the midst of the age of the corrupted, cultivated New Pagan selfhood.

The important thing to understand about the differences between Hofstadter and Penrose is that they are differences of faith. Hofstadter's materialism is circular: our brains work, they are physical systems; loops and paradoxes arise, but, since they are rooted in physical systems, they must have an end and this, in turn, is how our brains work. Physical systems, in other words, are physical systems, QED. Penrose says we do not know enough and, what we do know, suggests Hard AI is impossible. And he adds, in an explicit affirmation of faith, there is an underlying reality of numbers to which we have privileged access.

We may turn away from such an issue, bewildered and suspecting our own incompetence before Hofstadter's games and Penrose's equations. But this issue of faith is both far simpler and far more important than either man appears to be quite able to say. It is the ancient issue of who we are.

Science has not provided an answer, but it may do so on Hofstadter's terms. If it does we have an appalling problem. Science, as I have documented, has already stripped us of any number of versions of ourselves, leaving us with little more than a continuous state of uncertainty. In doing so it has reduced the human self morally and cosmically. For science now to discover that it can mechanically recreate or duplicate that self would precipitate an emergency. We would appear to 'know' ourselves in some ultimate sense just at the moment when we had decided there was barely anything worth knowing.

Yet, even without the sudden advent of Hard AI, we appear to be in crisis. We look upon these arguments and there is no way for us to choose between them. We have not the authority or confidence. Even if we cannot build a machine brain, where can we find the tools to rebuild ourselves?

9

The humbling of science

'If I have exhausted the justification, I have reached bedrock and my spade is turned. Then I am inclined to say: 'This is simply what I do.'

Wittgenstein[1]

Science made us, science broke us; it is time to start making repairs.

Let me first summarize the picture and the argument I have been trying to present:

A new and unprecedentedly effective form of knowledge and way of doing things appeared suddenly in Europe about 400 years ago. This is what we now know as science.

This science inspired a version of the universe, of the world and of man that was utterly opposed to all preceding versions. Most importantly, it denied man the possibility of finding an ultimate meaning and purpose for his life within the facts of the world. If there were such things as meanings and purposes, they must exist outside the universe describable by science.

This precipitated a philosophical crisis, first defined by Descartes. The crisis was one of knowledge. In a cold, meaningless universe, how did man know anything, science included? What guarantees of his knowledge could man find? More urgently the success of science also precipitated a religious crisis. As the physical evidence for religion was stripped away by successive generations of scientists, faith turned inwards to find a safe refuge inside the self. But this could not halt the progressive decline in the power of religion in the face of science's overwhelming effectiveness and scepticism.

In our century, however, science has been subject to new

doubts. Its more malign creations like the atom bomb have made people doubt its value and question its virtue. In environmentalism the progressive values of the society that science created have been rejected to be replaced by a new code of benign co-existence with nature.

But science itself has also changed. New developments have overthrown past certainties and many have come to argue that these developments show that science does not present a bleak, pessimistic vision. Rather it may lead us to a new spirituality.

Nevertheless, the project of hard science continues and still dominates our culture. Progressively science is becoming the culture of the entire world. Now its ambition is to unravel the workings of the human self.

That is the story of this book so far. It is underpinned by two convictions: first that the modern, liberal-democratic society has been created by the scientific method, insight and belief, and, secondly, that both this society and science itself are inadequate as explanations and guides for the human life.

Science now answers questions *as if* it were a religion and its obvious effectiveness means that these answers are believed to be the Truth – again *as if* it were a religion. But it confronts none of the spiritual issues of purpose and meaning. And, meanwhile, its growing power enables it to drive the very systems that did confront those issues to the margins of our concern and, ultimately, out of existence. As I have said before, our science, whatever it may pretend, is incapable of co-existence.

I have described many of the twentieth century's impassioned responses to this state of affairs and I will here run briefly through my reasons for rejecting them.

Environmentalism As a demand for housekeeping of the planet, this is no more than sound practice, assuming its analysis of the problem is correct. However, environmentalism has expanded to become an entire moral, social and political orthodoxy. As such it has joined forces with a whole range of other anti-progressive movements which advocate the abandonment of economic growth and the return to 'natural' ways of life.

The problem for me is that their conception of meaning and purpose is wholly negative. Our purpose is to survive by restoring some form of natural balance, and our meaning will then be implicit in that balance. But the purpose side of the ecological deal says only that we have an obligation to survive – scarcely a significant spiritual insight. In any case, how can merely undoing our ecological sins constitute anything more than a purely practical programme driven by necessity? We cannot be forced into a new spirituality merely because we have made a number of biological and chemical mistakes. The real revelation would only be that we were prisoners of an environment that wished to deny us the best of ourselves by effectively discarding the whole culture.

Finally, I do not believe there is any meaning or consolation to be found in simple advocacy of harmony with nature. All our history has demonstrated at least one, perhaps only one, indisputable fact about human life: we are fundamentally and irrevocably different from the rest of nature.

A return to orthodox religion Of course, orthodoxy has never quite died. It has survived either as fundamentalist faith or in various forms of liberal theology that attempted to redefine religion in line with the progress of science. Straight faith is either available to you or it is not. There is nothing more that can be said. On a wider level, however, it can be said, again, that science does not co-exist. Faith has been and will continue to be eroded by science – it may work for you, but the numbers for whom it will work will tend to decrease. As a political and moral force, therefore, it will be weakened.

Liberally redefining the faith to embrace or co-exist with science is unconvincing because it is too obviously trying to make the best of a bad job. There is no certainty of where such definitions should stop. Yet some degree of essential orthodoxy must be present to stop these new versions of faith becoming too hopelessly vague to work as religions. In other words you cannot have even a liberal orthodoxy without faith, so liberalism does not solve the problem of the need to believe. It merely attempts to pretend it is not a problem.

However, I should say here that this neither discredits nor invalidates religion as such. The state of affairs I am addressing is one in which religion has become less available to people. It is not one in which it has been shown to be 'untrue'. My own view might be said to represent a pre-religious argument for non-religious people.

A new spirituality of science   In one form this could – and does in writers like Bronowski, Sagan, Hawking, Feynman and Hofstadter – arise from our straightforward acceptance of the progressive, evolutionary vision that science provides. This says that the classical, scientific project is our destiny and we construct our spiritual identities in relation to that destiny.

This seems to me to be a passive submission rather than a way of confronting the issues I have raised. Its proposition is, in essence, that science is the truth, there is nothing we can do about it, so we might as well submit. Philosophers have colluded with this. Both Bertrand Russell and A. J. Ayer virtually condemned their subject to being the handmaiden of science. This was one of the great intellectual low points of our age. They reduced truth to effectiveness and assumed their job was done. I believe that it is self-evident that, if we are to have philosophy or religion, the first qualification of any claimant to those titles must be that they are different from and independent of science.

A new spirituality arising from within science   By this I mean the hope many have derived from modern developments like quantum mechanics and chaos theory. Some – like Fritjof Capra – say these point to a possible future convergence between ancient religious insights and new scientific ones. Others – like David Bohm – attempt to construct entirely new visions based on the anti-mechanistic tendencies of the new science.

But, as I have repeatedly said, science is mobile, its very nature is constant change. One generation's certainty is quite likely to be overthrown by the next. It may be true that quantum mechanics points to a deeper, spiritual realm – but the knowledge of that truth must come from outside and be independent of the quantum, otherwise it remains dependent on the whims of science. We must, in effect, know the truth before we can discover it in the quantum.

There is one further, general cause to be sceptical of all these responses. This reason is simply the strange fact of human life that a reason to live cannot be invented.

Understanding this is central to understanding the present, perhaps even the future, and certainly it is the key to an understanding of the failure – a magnificent failure, but still a failure – of the entire Enlightenment project. That project was to find a new basis for knowledge and value. It failed because, in essence, you cannot *think* your way to that basis. Max Weber captured the point brilliantly when he anatomized the quandary of the intellectual:

The need of literary, academic, or café-society intellectuals to include religious feelings in the inventory of their sources of impressions and sensations, and among their topics for discussion, has never yet given rise to a new religion. Nor can a religious renascence be generated by the need for authors to compose books, or by the far more effective need for clever publishers to sell such books. No matter how much the appearance of a widespread religious interest may be simulated, no new religion has ever resulted from the needs of intellectuals nor from their chatter.[2]

The wisdom and the limitless implications of that remark seem to me to reach into the heart of the modern predicament. The whole of the history I have described can be seen as a kind of footnote to Weber's insight. For that history is primarily an intellectual one, a history of ideas. Ideas, as I have throughout insisted, comment on what we are, they define the contours of our identity. But they cannot in themselves justify that identity because we cannot think our way to purpose and meaning.

A moment's self-examination will reveal how obviously true that is. We know in our daily lives that we either feel at one with ourselves or we do not. We know also that no amount of reasoning can change either state of affairs. A doctor might come up with some hormonal explanation of your condition; a psychoanalyst might suggest an incident in childhood. But the point is both would be offering only causes, they would not be explaining or describing the feeling itself. This is the same as the point I made about the colour green – knowing the particular wavelength of light does not tell us directly about the experience of green. These feelings of greenness or oneness happen and they exist in the world as surely as rivers, mountains and stars. That we think we can understand them through hormones, analysis or wavelengths says nothing at all about that existence, though it does say much about the kind of explanation we want to believe.

The whole point I am making is that a hard, irreducible sense of our own self-awareness has been progressively denied us by the inroads of science both as a form of truth and as a creator of our society. The chains of scientific causality have gradually appeared to explain away our moods and insights as surely as they have evicted God from the immediate neighbourhood of the solar system. In the past the sure sense of the self has always been sustained by religion in a variety of guises, indeed it might be said that this sense is the religious aspect of us all. But science has taken away our religion and we cannot invent a new one because religions are, by definition, not inventable. Certainly theologies can be invented, but, to be effective, they must spring from some deeper human well than the mind of a theologian; they must arise from a totality of human experience, they must be ourselves. And we cannot invent ourselves any more than we can invent the sensation of greenness.

For all these reasons, but, pre-eminently for this last one, I find all the attempts I have so far described to establish a new place for us in our world unconvincing.

At this point it might be tempting to say that, if we cannot *think* our way out of this problem, then none of this matters. Nothing can be done. In any case, most of us live out our lives with a reasonable degree of balance and sanity. If science works and scientific-liberal society is the most effective and just form of social and political organization yet devised, what more can there be to worry about?

The first answer is that not to worry about our spiritual impoverishment implies a shocking, passive, animal acceptance, a terrible inversion of human values.

The second, more practical, answer is that the situation in the scientific and affluent West is not stable. We cannot assume that our present spiritually inadequate but materially and politically successful societies will continue. And the primary reason for that is that they will increasingly find themselves unable to think of *reasons* to continue – precisely because the inhabitants of those societies are in the process of losing their sense of themselves and their culture. (This word 'culture', I know, is a problem here. See Glossary for my usage.) This is due to the progressive weakening of the unliberal, unscientific side of the society.

As a slight digression this might be seen as a denial of Francis Fukuyama's celebrated 'End of History' theory. He takes the view that science has imposed a progressive direction on history and the end of that progress is liberal democracy. But our differences are mainly of emphasis. For Fukuyama is obliged to acknowledge in the midst of his apparent triumphalism that scientific liberal democracy does not provide any final justification of the individual human life – indeed he admits that modern thought has arrived at an impasse in this area. For me this is a far more unstable state of affairs than it seems to be for Fukuyama.

Consider, for example, the kind of moral problems that science almost daily strews in the path of liberal society. Do we abort every embryo with an abnormality? Do we administer fatal doses of drugs to the dying? Should every organ of every corpse become available for transplantation?

These are the typical debating points that arise in Western societies about the demands of science. There are probably hundreds more to come over the next decades as the rate of scientific innovation continues to grow exponentially. Our liberal response to them, if they require legal control, is broadly this: we assemble a committee consisting of, typically, academics, clergymen and experts. They then come up with an answer. This answer will be a compromise between the two extremes of possibility. In the case of, for example, allowing scientists to experiment on human embryos the two extremes will be: the scientist can do what he likes as these embryos are not really human beings or the scientist can perform no experiment at all because these embryos are human and therefore sacrosanct. The compromise solution recommended by the committee on this issue in Britain was to allow experimentation up to the fourteenth day of foetal development. This, it was said, was when the 'primitive streak' appeared and the embryo ceased to look like a random clump of cells.

But this is, of course, no answer at all. The committee simply looked at the hard religious argument, then at the hard scientific and drew a convenient and temporarily convincing line between the two. And even that line was scientifically determined since it was the scientists who told the committee about the primitive streak. The only real answer to such an issue would be a statement of principle and a law embodying that principle. But, by definition, liberal society is too open-minded to have any such principle so it compromises with the competing demands of such a society.

I work as a newspaper columnist and I have examined many such issues as well as ones that appear, at first, quite different. I feel obliged to try to make up my own mind, so I argue at length with the protagonists. The pattern is always the same, no matter what the issue. Each side advances arguments arising from a basic conviction one way or another. But the arguments themselves are almost always irrelevant, a distraction designed to persuade but not really believed in as such. What is believed in is the basic conviction: either, in the embryo example, that scientists must be free or that human life at every stage is sacred. The beliefs will be held because of the irrational demands of temperament, upbringing and self-interest and they will, therefore, be absolutely irreconcilable. And, because they are irreconcilable and there is no external principle to which anybody can refer, the committee is incapable of answering the question it had been asked. So it simply draws a line, closer to the scientists or closer to the religious, largely on the basis of the political and social climate of the time.

Now, given the seductive effectiveness and persuasive power of science, over time it is clear that this line will tend to move further and further over to the scientific lobby. The pressure on the other side will be decreased as science continues to conquer because of its corrosive and restless refusal to co-exist. So I have two points here: all moral issues in a liberal society are intrinsically unresolvable and all such issues will progressively tend to be decided on the basis of a scientific version of the world and of values. In other words they will cease to be moral issues, they will become problems to be solved. The very idea of morality will be marginalized and, finally, destroyed.

The point is the weakening of the arguments against the scientific demands. The only real argument is a straight assertion of a faith shared by fewer and fewer people with less and less conviction. Because the philosophers of the Enlightenment failed to find another answer, there is no rational rhetoric that can compete with that of science. So those who feel uneasy about any embryo experimentation have nothing to say except 'We believe . . .' and in a liberal society that can only be understood as 'It is our opinion that . . .'

I believe this process is endemic in liberal societies. The progressive weakening of the non-scientific argument is matched by a weakening of non-scientific understanding. This process is so pervasive and so familiar a part of our lives that it is, in fact, quite difficult to grasp.

Take, for example, the idea of adolescence. In the 1950s it was established as a kind of institutional fact of life that teenagers rebelled against their parents in specific ways and this fact hardened in the 1960s into a form of political programme. This was all felt to be inevitable, a biological, sociological or historical necessity. The reality was, of course, that increasing affluence and the spread of mass communications were responsible for creating the styles of this highly formalized revolt. It was necessary for these children to be different in order that they would consume differently.

But any Marxist might say as much as that. What was far more important was the content of the adolescent revolt. In essence, this consisted of a glorification of change, of perpetual movement. The images are obvious. On the one hand there was the static, hidebound world of the suburban parents, restrictive and resistant to the freedom of the young. On the other were the young themselves: on the streets, on motorbikes or in cars, mobile, sexually free and, in the sixties, revelling in chemical transformations of consciousness. The one injunction was – and still is – to go with the flow, to be open and available to change. Parents might be portrayed as resisting this, but, in reality, they could not. The reason is the same as the reason we cannot conclude moral arguments – the Enlightenment left us with no fixed point from which the values of the culture could be defended. So, inevitably they will be eroded.

An adolescent act of defiance has now become a familiar form of virtue. 'Going with the flow' has become an imperative, a clear version of the good life. Change is sentimentally evoked in popular fictions. We 'change' to move on from tragedy or difficulty, we 'change' to overcome trauma, we are frequently told to accept 'change' as some perennial fact of life, the contemplation of which brings wisdom. Virtuous change is the ethic of the soap opera and the *ne plus ultra* of the advice column.

The idea of progress, of course, had been part of politics long before the idea of change had entered our emotional vocabulary. The American 'dream' or the European 'ideal' are both ways of saying that things must always be in movement forward. This movement is the purpose liberalism offers to individuals. What we do and attempt to achieve is forever targeted at some distant point in the future. With the institutionalization of the idea of change associated with adolescent revolt, the process enters human life – our state of being is forever provisional, always available to be changed into something else.

Time, in this context, takes on a specific moral dimension. Future time is good, past time bad. We move from this inadequate past into this bright future. Since progress is seen to be happening and is regarded as a virtue, the past comes to be understood as an underdeveloped realm, an impoverished Africa of memory and the imagination, useful only as a staging post for the future. Its significance is thinned out until it becomes a mere prologue for the present and the future. History is a dusty archive of doubtful value.

The most obviously inhuman aspect of all this – both progressive political insistence on forward movement as well as the cultural virtue of going with the flow – is that it denies the possibility of peace within human life. If there is only an eternal project of progress and a perpetual morality only of movement, then *my* life is of no value. If I accept these ideals, then I am destined to live and die for a cause that can never be triumphant. The point is that constant movement does not offer the individual a way of understanding his life other than as an episode in the flow. It does not offer a way of grasping and giving value to time in the only context that we genuinely know time – the context of the duration of our own lives. Yet, in spite of ourselves, we embrace this fiction of the flux.

This doctrine of undifferentiated forward movement can be illustrated by the way we understand and are taught about the great figures of our culture. If we are told about Abraham Lincoln or Isaac Newton, we are told about them as having done something that helped us to attain our current condition. Lincoln was instrumental in the creation of the modern American state and Newton led us to our physics and cosmologies. But, logically, this should mean that later American politicians or later physicists were greater than either. What we need is an ideal of greatness that refers to the quality of these men's lives, as individual lives, not as just visible landmarks in some interminable landscape of flux. Otherwise greatness will become meaningless. History will become an insignificant landscape of ages that were trying and failing to become our age. And, of course, in time, our age will be reduced to the same condition. Everything is reduced by the idea of progress and change.

Such a climate and such attitudes are the climate and attitudes of science. Science created the idea of constant forward movement and of the possibility of complete transformation. Its experimental methods and attitudes are implicit in the revolt of the young. And, of course, its ruthless rejection of the past is implicit in the impotence of the parents. For the truths of science do not require the wisdom of the past. A computer scientist need never have heard of Newton, he need only know of a thin film of recent knowledge to be able to master his art. Scientific progress is so radical that, at every stage, it is able to throw away almost all the baggage of its own history.

This has profound and utterly negative spiritual implications for the inheritors of the scientific legacy. The America academic Allan Bloom has brilliantly and movingly anatomized these implications in his book *The Closing of the American Mind*. Bloom was inspired to write by his increasing dismay at the spectacle of American college students. They appeared to be progressively more lifeless and ignorant. But, even worse, they were entirely unable to view their own culture as anything worth defending. A terrible cultural relativism had invaded their lives, denying them the possibility of choosing one point of view as more valuable than another.

'The study of history', Bloom writes, 'and of culture teaches that all the world was mad in the past; men always thought they were right, and that led to wars, persecutions, slavery, xenophobia, racism and chauvinism. The point is not to correct the mistakes and really be right; rather it is not to think you are right at all.'[3]

Bloom had identified the way in which a key liberal attitude had penetrated and corrupted the student mind: people thought they were right in the past and did terrible things as a result; so we must never believe we are right.

Science and liberalism will not give us the means to defend what we, in particular, are because it will not acknowledge the possibility that we, in particular, are right. So the justifications of the parents, the teachers, the entire culture sound ever more hollow. Apparently there is nothing to teach since any new development may invalidate any old fragment of knowledge. Certainly there is no cultural core, no body of virtue to be transmitted. Our souls become enfeebled. Knowing nothing and thinking nothing, we wander through life as through a bewildering, undifferentiated freak show. Why is this saint more important than this bearded lady? What does this philosopher know that this clown does not? Who am *I* to decide? Who are *you*?

This fluidity, this refusal of all hierarchies is now thoroughly institutionalized. On American campuses there is the concept of 'PC' – politically correct – in which the slightest suggestion that the European-American way of life is superior to another is rooted out and forbidden. Students are simply not allowed to value the culture that made them more than any other. And, if they cannot do that, they cannot be taught.

Of course, this in turn produces the perverse phenomenon of liberal authoritarianism. Precisely because liberalism finds itself with so little to say, it says what it can with ever more illiberal conviction. Certain companies, for example, now indoctrinate their employees with green, liberal values with a ferocity almost indistinguishable from the most severe religious or political sect.

The inhumanity of the idea is flagrant. People live their lives by making distinctions of value. They prefer their families and friends to complete strangers, they prefer this town to that one and so on. But value distinctions are not allowed, so people must not live like people.

Bloom's culprit is the same as mine: 'Science, in freeing men, destroys the natural condition that makes them human. Hence, for the first time in history, there is the possibility of a tyranny grounded not on ignorance, but on science.'[4]

His point is that, so enthusiastic has science made us for

rooting out and destroying delusions that we identify this activity with education and enlightenment. Our wisdom has become entirely negative. All that we pass on to our children is the conviction that nothing is true, final or enduring, including the culture from which they sprang. Knowledge is science, all else is speculation, fantasy or wishful thinking. But this must be wrong.

'Think of all we learn about that world from men's belief in Santa Clauses,' writes Bloom, 'and all that we learn about the soul from those who believe in them. By contrast, merely methodological excision from the soul of the imagination that projects Gods and heroes onto the wall of the cave does not promote knowledge of the soul; it only lobotomizes it, cripples its powers.'[5]

Of course, the liberal response is that Santa Claus is not 'true' so he cannot constitute any form of knowledge. He is a delusion and rooting out such things is at the very heart of the liberal-scientific project.

'One of the greatest benefits that science confers upon those who understand its spirit,' wrote Bertrand Russell, 'is that it enables them to live without the delusive support of subjective certainty. That is why science cannot favour persecution.'[6]

I treasure this remark as being one of the most fabulously stupid by one of the most fabulously stupid men of our age. It embodies the liberal creed in its most abject form as messenger boy for science. It is the impotent creed of Bloom's American undergraduate expressed in the trappings of high philosophy. What Russell is actually saying is what those students are living: people who thought they were right in the past have done terrible things, so we must never think we are right again unless we are doing science because that will not lead to us doing terrible things. The idea is craven and inhuman. It demands that we disregard everything that is truly ourselves – everything that is utterly real to us, though outside the reach of science – as just a dangerous mass of delusions, of uncertainties with no external value. It never occurs to Russell to consider that it is the terrible things that should be eliminated rather than the 'subjective certainty'.

This is truly the corrupted fag-end of the history I have been describing. What began with the hard sanity of Galileo, the visionary genius of Newton, the soulful perplexity of Descartes and the magnificence of Kant ends with the spectacle of the vapidly pragmatic Russell, subjugating philosophy to science, sneering at faith, tossing aside the human imagination and abandoning the last traces of an idea of morality. If my story has to have a villain, let it be Bertrand Russell.

All of this excursion through the morality and culture of liberalism was, you will recall, in answer to the question: What more is there to worry about? My first answer was that it is inhuman not to worry about our spiritual impoverishment. My second answer is that liberal society is unstable and in deterioration. The scientific understanding as a basis for human life is radically inadequate, yet it continues to triumph. As a result, human life itself will become inadequate. That is what there is to worry about.

Santa Claus, however, does not exist scientifically and we cannot reason him back into such an existence. A reason to live, I repeat cannot be invented. It is all very well wanting to believe, but we don't. I will now attempt to describe an alternative. It is not another invention because it is already within us; indeed, it is us. And it does not require a programme of action or mass conversions because it is almost certainly already happening.

As a prelude to my alternative, let me repeat a point I made earlier. People live unliberal lives. They have values, convictions, preferences and loyalties by which they order their world and make it work.

As I write this there are news stories about the possibility of the release of Western hostages by terrorists in the Lebanon. A Briton, John McCarthy, has just been released. There is a mood of wild, national happiness. Somebody – a liberal – makes a point that this is irrational and selfish in view of the hundreds of Arab hostages in Israel or the thousands unjustly held elsewhere. The reason this point is foolish is that McCarthy is one of our own and, therefore, we value him more. There is nothing wrong with this.

It is, I believe, humanly impossible actually to be a liberal. Society may advocate liberal tolerance and open-mindedness, but nobody practises it. In fact, this is what preserves liberal society.

For a complete personal acceptance of scientific-liberalism would reduce the society to passive, bestial anarchy. There would be no reason to do anything, no decisions worth making and certainly no point in defending one position as opposed to another. What I am saying is that we already *live* the solution, even if we do not know it. In this sense science's invasion of our souls is incomplete and may always be incomplete.

Understanding the nature of my solution requires me to turn to what may, at first, seem like a philosophical technicality. Ludwig Wittgenstein died in 1951. Two years later his book *Philosophical Investigations* was published. Wittgenstein had been a pupil of Russell's, but they had drifted far apart. Russell was to describe Wittgenstein's later philosophy, embodied in this posthumous book, as 'incomprehensible'. Being incomprehensible to Russell is, of course, my personal touchstone of the highest virtue.

The book, among many other things, contains a celebrated argument about the idea of private languages. A private language would be one that only had meaning to the user. The example is employed of a man who wishes to record the experience of a particular sensation in his diary. It is not a pain or an itch, there is no word that describes it. So he uses the letter 'S' to record each occurrence of this sensation. Now this letter 'S' might be taken to be a word in a private language that has meaning only to the man. But Wittgenstein concludes that it is not, rather it is quite meaningless. The point is that, in order to get to the word 'S', the man had to go through the language we all use. To say that 'S' stood for a sensation requires him to employ the word 'sensation'. He cannot isolate himself and his words from the public realm of language. He must have language before he can have a sensation. There cannot be such a thing as a private language because language is, by definition, a public thing.

As I said, this may seem to be a technical point. But place it alongside Descartes and its profound significance begins to emerge. Descartes's *cogito, ergo sum* was an assertion that the one thing of which he could be certain was his own experience on the basis of his own thinking process. But that is an assertion of a private language. It depends upon the idea that we can address words to ourselves, absolutely independently of the outside world. The absolute independence is essential if Descartes's argument is to make any sense at all, because, if it is dependent on the outside world, then he has established nothing, for his own thought will have been revealed to be part of the phenomenon of existence towards which he is committed to an attitude of complete scepticism. But Wittgenstein destroys the point: the *cogito* is like 'S', and we cannot arrive at it without going first through the public realm of language.

As we have seen, the Cartesian self was central to the imaginative power of science. The idea of this isolated, thinking thing, trapped in yet separate from the body, has been the foundation of the modern conception of the world. Stripped of Descartes's God – his bridge back to the reality of the world – it has been a kind of modern bottom line. We have assumed that the absolute loneliness of the Enlightenment selfhood is both true of our human condition and necessary for the effectiveness of our knowledge.

But it is neither. It is an act of faith and, as Wittgenstein shows, it is an incoherent act of faith. It finds in language an absolute, internal basis for the self, whereas what we actually find in language is an absolute, external basis. Language comes before the *cogito*; language gives us our selves.

Locked in this remote and difficult philosophical work this is, I believe, the first – and most entrancingly beautiful – sign that we might be in the process of escaping from the loneliness of the classical scientific vision. For, if the scientific self is revealed to be a convention, a delusion Russell would say, then it follows that science too is a convention, a specific choice rather than the privileged road to the truth. Science may, at last, be relativized and thereby humbled.

Wittgenstein's insight is philosophically technical, but it is simple. It is not so much an idea as a correction of a misconception that is embedded in our view of the world. Indeed, his whole view of philosophy was not that it came up with ideas, theories or innovations, but rather that it was a process that allowed us to untie the knots that had formed in our minds. He saw the way science and logic had tied most of these knots to create a curious and baffling barrier between ourselves and the world.

They make us, he wrote, 'want to *understand* something that is already in plain view. For *this* is what we seem in some sense not to understand.'[7]

By untying Descartes Wittgenstein dealt with one of the most intractable of these knots. He was not inventing anything, he was undoing the damage that has been done. One cannot emerge from Wittgenstein's insight saying: '*This* is what I now believe.' One can only 'see' the ways in which we have constructed unnecessary problems for ourselves. Perhaps – and I will admit this much of the science-based metaphysical speculation into my argument – this 'seeing' is the same as the 'seeing' involved when a mathematician like Professor Roger Penrose 'sees' a truth without being able to prove it. It may be that Penrose is right – this irrational, inexplicable, non-computable ability of ours to 'see' things lies at the heart of what it is to be human.

But Wittgenstein was too engrossed and intense a man to speak directly of the implications of his later work. He did not address the world as a prophet. And this, too, has its significance. For he saw that, in fact, there was nothing to say, that one could only record the fact that the knots had formed and the way they might be untangled. And, when they were untangled, one could achieve a degree of peace with oneself.

The American philosopher Stanley Cavell summarized this idea: 'The more one learns, so to speak, the hang of oneself, and mounts one's problems, the less one is able to *say* what one has learned; not because you have *forgotten* what it was, but because nothing you said would seem like an answer or a solution: there is no longer any question or problem which your words would match.'[8]

I believe this wisdom to be the most profound of our age, but I see that it might appear, in this form, remote from the problems I have undertaken to solve – the problems, in essence, of how one might arrive at a post-scientific society. Given the potential decay of liberalism that I have described, these problems must be practical and urgent. They cannot obviously be solved by philosophers.

But Wittgenstein's insight is that there is no problem at all. At least, there is no problem in the sense of something that requires some sort of innovation that we call a solution. Rather there is a difficulty, a blockage in our understanding of the world which can be removed not by a solution but by a realization. In this sense he is saying that it is the philosophers who are wrong rather than the people who simply get on with their lives. By thinking in terms of problems and solutions and by being hypnotized, like Russell, by the scientific conception of truth, they have tied the knots. And the knots have turned out to be contagious. Because of the success of the science the philosophers have defined, the knots are now in all our minds. In the twentieth century science and philosophy have formed an unholy alliance to defeat and discourage our sense of ourselves.

Let us return to the realization that language precedes the Cartesian *cogito*. This does not simply mean that we must, therefore, turn our attention to an analysis of language, as many have done, as the most important area of scientific inquiry. That is merely to tie new knots and they can be seen everywhere in the contortions of recent forms of thought like structuralism, post-structuralism, post-modernism and so on. These deliberately hermetic intellectualisms certainly arise from a contemporary realization of the primacy of language. But the response is no more than an attempt to draw language into science. Language becomes the undifferentiated flux of movement into which we are born. Deciphering the movements of this flux becomes the new project of science.

But this is an astonishingly banal response which utterly ignores the ethical and spiritual significance of the overthrow of the Cartesian self. For the language of which Wittgenstein writes is not simply some objective 'stuff' which we can dissect as we might an exotic organism or observe as we might the electromagnetic spectrum of a star. Nor is it even the particular grammatical structures that we employ when we write or speak. Rather it is the very nature of our consciousness – the very nature of our *self-consciousness*.

This realization does not leave us drowning in some fashionable post-modernist flux; rather it lifts us completely out of the flux and replaces us in a world made of the real experience of our individual lives. Our self is real and it is given to us by the whole of language, which means by the whole of culture and history. We are defined by not less than the entire culture. Wittgenstein pointed out that to understand one note of Beethoven, we needed to be able somehow to be connected to the entire culture from which it sprang. We are tied into our history and our way of life because it is all embodied in our language – indeed, that is the only place that it can be said to exist – and we are formed out of language. We are given selves that are made up of everything that has ever happened, we are bathed in and embodiments of history. And this history includes magic and religion as surely as it includes science. The colour green is real; feeling at one with ourselves is real. Wavelengths, medicine or analysis can only come after these facts. Science's trick was to pretend it could tell the story of what happened *before*.

A recent, brilliant, philosophical attempt to move on from Wittgenstein and unite him with a positive, materialistic conception of the self was Daniel C. Dennett's *Consciousness Explained*. Dennett provides a comprehensive, materialistic but strictly anti-Cartesian conception of the human mind which supports Douglas Hofstadter's conviction that there is no intrinsic reason why artificial intelligence is impossible. Dennett arrives at a brief moral conclusion that is broadly similar to mine – that the culture and our selves as they are must be of an undeniable and final value that cannot be argued or rationalized away. But, he fatally adds, things change and we cannot cling on to the conviction that, because things matter now, they will always matter.

Again there is the frustrating incompletion. Again there is the sense that something vast is being subtracted from the content of the individual life. Again there is the sense that we are moving forward in a process of progress disillusionment. However intriguing and liberating Dennett's conception of mind may be, it does not lead us out of the crisis of the Enlightenment.

In contrast:

'A instinctive vision of health and peace underlies our horror stories,' writes the American novelist John Updike. 'Existence itself does not feel horrible; it feels like an ecstasy, rather, which we only have to be still to experience.'[9]

Updike is a writer at the heart of the Protestant tradition and his words here spring from that tradition – from the tradition of Kant as well as Kierkegaard that insisted on the primacy and irreducibility of the inner human experience. This emphasis suggests the moral and experiential form of the insight I am attempting to convey. They were right, they saw the limits of science. They saw that something real – an ecstasy in Updike's terms – was in the world which it was pointless to deny. Kant turned that insight into a vast, metaphysical system, Kierkegaard into the foundation of authentic choice. But they were confronting the same problems, the problems of the scientific, Cartesian universe. What Wittgenstein does and, implicitly, what Updike is expressing, is show that there is no problem to be confronted. Our inner ecstasy, our sense of ourselves is not to be opposed to, and thereby be potentially defeated by, the vision of science. Rather we have to see that science springs from the same source, it is *part* of us, and only one part. Perhaps the modern form of authentic choice is simply the choice to stop doubting ourselves.

In a discussion with the biologist Richard Dawkins I asked him about the Cartesian view that animals were machines. Descartes had excluded people because his God ensured they were more than machines. But Dawkins asked simply how we could define a being that was not a machine – as, he suggested, something with an immortal soul. Because he did not believe in any such thing, people were machines. Dawkins has sound and consistent reasons for believing this. But it is a choice and, to me, it is incoherent. For people throughout history have felt they have souls. This feeling is real, it is not modified by either psychoanalysis or physics. Indeed, the word 'machine' only functions in the language at all if we do exclude people – it means an artificial mechanism that is not a person or it means nothing. This is not a quibble, it is at the centre of what we are. The way the language works tells us at every moment how to be ourselves. We are not machines because the language tells us so. Perhaps from that we can conclude we have immortal souls. All right, human beings as a whole may have invented or evolved the idea of a soul. But does that make it any less real, less permanent? It has outlasted all scientific conception.

So Allan Bloom was more right than even he knew when he raised his point about Santa Claus. Of course, the liberal-scientific establishment will think it has achieved something each time it teaches somebody that he does not exist. But, Bloom says, think what Santa Claus teaches us about the human soul. True. But there is an even more powerful truth – Santa Claus does exist.

He exists unscientifically. I could, of course, make up some imaginary monster now and say it exists in the same way. But the point about Santa is that he has *more* existence precisely because of the way he is embodied in the language of all who have ever believed in or even just heard of him. To be a dreamed-of dragon is one form of existence, to be a believed-in Father Christmas is another, higher form.

I stress again that all this is a realization, not a new ideology, metaphysic or opinion. It is a change of emphasis that springs from the discovery that we no longer have to subject ourselves to the suffocating demands of the scientific sense of ourselves.

Seen from this perspective much of what I have been describing can be seen as symptoms of this change. The Anthropic Principle, for example, demonstrates an important doubt and the necessity of a vital modification of the classical scientific definition of our knowledge. Equally, the need for, if not the facts of, many of the attempts to find spiritual sustenance in the new forms of science indicate a drawing back from the heroism formerly attached to the scientific quest. Environmentalism springs from the same impulse but adopts a more nihilistic attitude towards the culture which created science. It is an understandable response but one which fails to take into account the values embodied in the *whole* of that culture. And remember how, in Chapter Seven, I spoke of the downgrading of technology as an exciting end in itself, how the neutral boxes of the computer signalled that the idea of a mechanical Utopia was losing its power even at the level of popular culture.

All such symptoms begin to make it clear how dated are the popularizations of men like Hawking, Sagan and Bronowski and how iniquitous are the efforts of men like Russell to provide apologias for science. The popularizers are popular because they make sense of all the difficult science we are dimly aware is going on about us. But we think that this science will provide more than just another hollow, mechanistic vision. So, when we read of the strange contortions of time, the fabulous eccentricity of the universe or the strange play of light on matter, we are drawn, with relief and excitement, into the idea that the world is more extraordinary than our reduced, modern idea of our lives had allowed.

But this is a trap. For these writers are, in reality, selling no more than the hollow mechanistic vision that made the familiarly reduced version of ourselves and now threatens to destroy us. They will put us back on the treadmill of eternal progress, reducing the span of our lives to trivial, accidental interludes and removing our souls far from our bodies.

But the story they are telling is only one story. There are many others, but one is of particular importance. It is a secret one that has been told in a whisper in the few dark shadows left in the bright glare of the history of the Enlightenment. The heroes of this story are men like Pascal, Kant, Kierkegaard and Wittgenstein. Intellectual heroes, certainly, but heroes who were moving towards a realization that was anything but intellectual. And they are the heroes who signalled the historical position in which we now find ourselves – at the end of the Scientific Enlightenment.

That this end is only just beginning must be apparent from the story I have told. Science is still triumphant and our liberal societies are still scientific. But we are clearly in a decadent phase and, I think, a terminal one. The decadence arises from the obvious failure of liberalism to transmit any value other than bland tolerance. It cannot defend itself and it cannot celebrate itself. Education is in permanent crisis because neither pupils nor teachers have any faith in what is to be taught. Bloom's portrait of American students is memorable and powerful because it is familiar. Teenagers have been taught plurality and tolerance and, as a result, think it is a virtue not to make up their minds about anything. They become blank, deadened inhabitants of the flow because they conceive of the flow as virtuous. The 'road movie' is a sacred text. Occasionally discontented, they may pick up Hawking or Bronowski in a dim aspiration towards metaphysical succour. But they find only propaganda, the dissemination of unease, the old scientific sleight of hand that makes effectiveness seem like truth.

I think – I hope – this suggests decadence to enough imaginations to make a change inevitable. This change involves not the reception of an argument but a transformation of attitudes, the first effect of which would be the relativizing, the humbling of science. Science would come to be seen as what it is – a form of mysticism that proves peculiarly fertile in setting itself problems which only it can solve. It does so by the employment of all the bizarre repertoire of thought processes I have described in this book. Its effectiveness is almost inevitable because it narrows the possibility of refutation and failure.

Science begins by saying it can answer only *this* kind of question and ends by claiming that *these* are the only questions that can be asked. Once the implications and shallowness of this trick are realized, fully realized, science will be humbled and we shall be free to celebrate our selves again.

And that should mean that science can become itself again rather than the quasi-religious repository of all our faith defined by the popularizers. We would have forced science to co-exist by turning it into something else, something more human.

There is a further important sign of the terminal decadence of scientific liberalism. Liberal thought had long sought a specifically liberal definition of values and virtue. All have failed because the idea is clearly impossible – how could plurality and tolerance alone provide a basis for concepts like 'justice' or 'the good'. Now much social and political thinking has now veered away from this idea towards an Aristotelian concept of 'the good' as the realization of specific excellence within a social context. This is an anti-pluralistic, anti-egalitarian, indeed, anti-tolerant attitude since it demands that choices be made between activities regarded as good in themselves and also that distinctions be made between people's behaviour in those areas irrespective of their private reasons for that behaviour.

For ourselves we can begin to define our lives in the terms in which we do anyway when left to our own devices. We can have irreducible affections, values and convictions. At their most fundamental these need not be defended further because they will express only our kinship with our culture and that kinship will be beyond appeal.

'If I have exhausted the justifications,' wrote Wittgenstein, 'I have reached bedrock and my spade is turned. Then I am inclined to say: "This is simply what I do." '

At the other end of the history of the Enlightenment, there was a prefiguring echo of this insight in the mind of Pascal: 'The strength of a man's virtue must not be measured by his efforts, but by his ordinary life.'[10]

What we are is what we ordinarily are. This is what we do. We are our own embodiment. If you are possessed by the suspicion at this point that I have told you nothing that you did not know already, that is precisely my point.

I am born and I shall die and, in between, these visions are what they most obviously are: mine. This is the only timespan I have and the only one in which my virtue and purpose may be found. I choose not to be written into some history of the future or beguiled by the technological demands of the as-yet-unborn. This is not selfishness, it is the ultimate unselfishness because it means I know what myself is – an expression and creation of my culture, a culture that has come close to sacrificing itself on the altar of one small aspect of itself. But I owe myself to *all* that culture and it must clearly be defended with my life because it is my life.

Such an avowal means the end of the rule of science because it denies the infinite open-endedness and willingness to change that science needs for its continued invasion of our souls. It also means an insistence that my soul be put back where it belongs – in my body – rather than in the remote realm to which, 400 years ago, science consigned it. This realization alone may not make that soul immortal, nor will it promise me an afterlife or salvation. So you may say it leaves me exactly where I was before – mortal, suffering and as lost as ever. I will reply that there is one vital difference: I shall not be, at the last, alone.

Glossary

I provide this glossary partly for the usual reason that certain words or phrases in *Understanding the Present* may require readily accessible explanation. But there seems to me to be an even more important reason. Words have been colonized by experts. They have developed specific meanings in specific areas. 'Liberalism', for example, can have six or seven different meanings depending on the context in which it is used. 'Classical' has almost been diluted out of existence: for some it refers only to the civilization of ancient Greece and Rome, for others to any music that is not pop and some see it merely as a synonym for 'timeless' or beyond the demands of mere fashion.

In normal use we tend to pick up on the likeliest meaning suggested by the context. In the realm of ideas this leads to a hopeless confusion. Philosophers, theologians, scientists and the man-in-the-street frequently use the same words to mean utterly different things. Often books that aspire to popularize complex ideas fail precisely because they omit to explain the fact that words that look and sound like the same ones we use in conversation actually have a far more specific technical meaning.

Clearly it would be absurd to suggest, to select and insist upon one definition as final. But, equally, it would be hopeless if a word evoked one notion in the reader's mind and a quite different one in mine. The list that follows, therefore, provides a guide to what I intend in this book.

Anthropic principle A principle in physics, defined over the past thirty years, which stresses the importance of the conscious, human perspective. In its weak form this says simply that our observations and theories must take into account the fact that we are here. The universe must have lasted long enough for conscious, carbon-based life forms to have evolved so the results of our observations of its present condition must be conditioned by the passage of that specific length of time. In its strong form the principle says that all the astonishing coincidences of physics, chemistry and biology that have conspired to produce us indicate that the fact that conscious life has evolved is the central, unique fact about this universe. This is, ultimately, what *defines* our universe. There is a Final Anthropic Principle which extends this even further to forecast that, once conscious life has evolved it will inevitably take over all matter. The important point about the Anthropic Principle in all its forms is that it potentially challenges the classical (*see below*) view that we can only understand the universe by objective observation of its workings irrespective of our presence (*see also* Fridge-Light Hypothesis).

Aristotelianism The thought of Aristotle (384–322 BC), the most influential philosopher and scientist of antiquity. This ranges over philosophy, ethics, cosmology and physics. In the case of the latter two disciplines his was the system overthrown by the Galilean-Newtonian revolution of the seventeenth century. It had been endorsed and united with Christianity by St Thomas Aquinas (*see* Thomism).

Artificial intelligence Often used commercially to define specific advances in computer technology. We are often told of computers possessed of AI which, on closer examination, turn out to be simply faster or more complex than previous machines. This seems to me to be a feeble usage and an abuse of the word 'intelligence'. The phrase is used here solely to evoke a possible machine that could reasonably be compared to a human mind.

Authentic choice A phrase derived from the theology of Sören Kierkegaard (1813–55). For him the decision to become a Christian was not one inspired by reason or persuasion. One did not have faith because Christianity was likely, rather one chose to have a faith. The authenticity of this choice arises from the fact that no arguments are being balanced, no advantages weighed up. We are simply being asked to choose and we do so. The importance of this is that it sidesteps the invasion of religion by science. Merely because science does not find a heaven 'up there', a hell 'down there' or does not detect the immortal soul leaving the body at death is irrelevant. Such 'reasonable' demonstrations are irrelevant in the context of the individual soul. We choose in the absence of all knowledge or evidence because that choice is what it is to be human.

Big bang The moment, predicted by relativity, at which space and time began.

Big crunch The moment, also predicted by relativity, at which space and time will end.

Black hole Again a product of relativity. These would be bodies whose mass had become so great that nothing, not even light, could escape the pull of their gravity. Since this would produce some very bizarre events as well as an interior of the Black Hole in which the laws of physics were suspended, these objects have been a fertile ground for wild speculation both within physics and without. No black hole has yet been found with absolute certainty.

Calculus A mathematical technique devised by Isaac Newton (1642–1727) and, apparently at the same time, by Gottfried Leibniz (1646–1716) in Germany. Its importance was that it provided an extension in the power of mathematics just when it was required to deal with a parallel extension in the power of physics.

Cartesianism Relating to the thought of René Descartes (1596–1650). Primarily used to describe his dualism – the division of mind and body – as well as his rigorous scepticism towards the evidence of the senses.

Causality The nature of cause and effect. Apparently self-evident in that a ball moves if we kick it – i.e. the kick causes the movement. But, in fact, different types and interpretations of causality can be used to define fundamental differences in our understanding of the world.

Categorical imperative A term of Immanuel Kant's (1724–1804). It describes a moral necessity that must, according to his philosophy, arise *a priori* – prior to the evidence of the senses – in the reason of man. A certain kind of action is objectively necessary because it is, as it were, a law of our existence. We must act out of duty and as if our action were at once to become a law of nature.

Chaos/chaology A development in mathematics based primarily on the realization that small changes in initial conditions can produce chaotic and fundamentally unpredictable changes in later conditions. The idea is still in its relatively early stages but has implications for all types of science that attempt to model conditions in the real world – weather forecasting, for example. As with quantum mechanics (*see below*) Chaos Theory suggests there is a real limit to our ability to know the world.

Classicism Classical science is sometimes said to be all science up to the year 1900. This is intended to differentiate essentially Newtonian physics from later developments like quantum theory and relativity. In fact, this is unsatisfactory since many later scientists would describe themselves as classicists, insisting that there is nothing essentially unclassical in, for example, quantum theory. To cover this I use classicism to refer to the belief that there is an objectively real world to which our scientific method gives us potentially conclusive access.

Cogito Shorthand term for Descartes's 'Cogito, ergo sum' – I think, therefore I am.

Copernican Relating to the thought of Copernicus, the latinized form of the name of the Polish astronomer Nicolas Koppernik (1473–1543). Almost always used to mean the belief that the sun, as opposed to the earth, is at the centre of the universe.

Cosmological constant Number invented by Albert Einstein (1879–1955). He saw that his theories would predict an unstable universe that would initially expand and then contract as a result of gravity. The cosmological constant was a force working in the opposite direction to gravity that would keep the universe stable. This was Einstein's real god.

Cosmology The study of the universe. I use in the sense of having competing cosmologies – say, Aristotelian cosmology as against Copernican.

Culture In some ways a disastrous word in that it tends towards a fatal vagueness in use. Typically we tend to use culture in a way that suggests cultural relativism – this culture as compared to that one and so on. As such the very word tends to diminish our particular understanding of the world. However, it is irreplaceable. My use is also general but is intended to suggest the totality of our ways of life and knowledge. This totality I regard as closer to the poet T. S. Eliot's idea of culture as an incarnation of a nation's religion than to the liberal idea of culture as a chosen way of life.

Design, argument from The argument for the existence of God based on the perception of the organized complexity of creation. A device so complicated and so well adapted, runs the reasoning, can only have been consciously made.

Determinism In human terms this is the belief that we are not free but subject to external forces over which we have no control. In terms of nature it is the belief that, since all effects have causes and these can be traced backwards in time, then the entire history of the universe was fully determined at its inception. Determinism is a logical attribute of classical systems.

DNA Deoxyribonucleic acid. The double-helix-shaped molecule which acts as a genetic message carrier.

Empiricism *See* Rationalism.

Epistemology The study of the way in which or on the basis of which we know things.

Existentialism A twentieth-century philosophy arising directly from the theology of Kierkegaard (*see* Authentic Choice). It insists on the individual as free and self-defining, morally obliged to be beyond the reach of all other influences.

Field theories A field in physics is a region in which a body is subject to a force arising from the presence of other bodies. A field theory is thus primarily a theory about such interactions. But their importance stems from Einstein's search for a 'unified field theory'. His point was that all reality was subject to fields and, therefore, such a theory would represent a kind of conclusion of physics.

Fridge-light hypothesis When we open the fridge door we see a light on inside. We may conclude that there is always a light on inside because every time we open the fridge door there is, indeed, a light on inside. This is a reasonable view of reality, but we know it is incomplete. For we know that the opening of the door turns on the light. This is a problem not of observation, but of perspective. We cannot fully understand this state of affairs without taking into account our role in the observation (*see* Anthropic Principle and Quantum Theory).

Geocentrism *See* Heliocentrism.

God He can be defined minimally as a conscious being that is not human or maximally as the spirit of all that is. He can be exclusively within you or exclusively without. We could say, vaguely, that He is the non-scientific cause or basis of creation; or, precisely, that He is the God of the Bible. For me His most important role is as the embodiment of the idea that human knowledge is intrinsically limited, that other knowledge is possible, but not for us.

Gödel's incompleteness theorem A crucial twentieth-century mathematical proof that no arithmetical systems can ever be complete. In terms of mathematics this overthrows Formalism and, in computer theory, it suggests we might never be able to construct a 'thinking' machine.

Heliocentrism The belief that the sun is at the centre of the universe. This belief was attributed to Copernicus as he found it to be more in accord with observation than the preceding belief that the earth was at the centre (*see* Geocentrism). Science, now, of course, does not believe either – merely that Copernicus was more right because the sun is now taken to be at the centre of the solar system. The emotional and philosophical aspect is more important than the astronomical. If the earth was not in the centre, our position in the universe was not special or privileged.

Idealism In common use, usually in politics, to denote attachment to a particular abstraction or view of human excellence. Philosophically it refers to belief of Plato that there exists a realm of ideal forms. These forms constitute reality whereas the images of our sense are but the distorted impressions of this reality. In Kant this becomes the conviction that we can have no direct knowledge of reality, only the play of the senses.

Liberalism I use this protean word specifically to refer to the form of society in which government's only real concern is the maintenance of order, plurality and tolerance. It can have no views on the ethical or transcendent nature of human life, but rather attempts to provide a stable, neutral arena in which competing convictions about such things can co-exist. But the logic of liberalism means that the liberal himself can have no such convictions. So the ultimate liberal society is one in which all such convictions have been eradicated. I take this to be the logical form of the scientific society and I also take it to be unlivable, meaningless and inhuman. Liberalism is the precise correlative of the scientific view that we must remove ourselves from the world in order to understand it – in liberalism the equivalent concept is that we must remove ourselves from values in order to understand them.

Newtonian mechanics The physical system defined by Isaac Newton based primarily upon his laws of motion and his gravitational constant. Its place in our picture of the world is so central that it might be said that the model provided by Newtonian mechanics might now be said to constitute our 'common sense' view of the world.

Physics From our schooldays we have come to think of physics as just one science among others. It is not; it is *the* science. Strictly speaking it is the study of matter and energy and their interactions. But this logically means that it is the study of everything. Clearly physics can be shown to underpin all other sciences. Biology can be seen as based on chemistry at the level of cellular interactions and chemistry can equally be seen as based on physics at the level of molecular and particle interactions. On the large scale the laws of physics – of gravity, of motion, of energy and of time – can be taken to be the determining factors of all other interactions. One way of understanding the Galilean and Newtonian revolution is as the moment when physics displaced theology as the Queen of the Sciences. Before, below, beneath, behind and beyond everything is physics. To the scientific imagination the world is made entirely and only of physics.

Psychoanalysis I use the word to mean the procedure invented by Sigmund Freud (1856–1939). Clearly there were other originators of psychoanalysis. But they all share the common strand of the attempt to evolve a science of the self based upon a causal narrative of the development of the psyche. The unravelling of this secret – because unconscious – narrative would, it was assumed, have therapeutic benefits.

Quantum theory/mechanics The body of scientific knowledge that sprang from Max Planck's calculation of the quantum of action – written as 'h'. This showed that there were only certain possible energy states and therefore nature was fundamentally discontinuous. Dozens of other bizarre and improbable implications follow that undermine the Newtonian picture of a universe of billiard balls interacting more or less according to the rules of what we have come to regard as common sense. So strange and unclassical are the products of quantum theory that Einstein spent the latter part of his life trying to prove them wrong or incomplete. But the theory still stands as probably the most practically effective ever.

Rationalism Commonly used simply to mean the belief in reason. Philosophically, however, it means that reason can contribute to our knowledge of the world either prior to or over and above the knowledge supplied by our senses. The belief that our senses provide the only knowledge is Empiricism. Bacon (1561–1626) was an empiricist, Descartes a rationalist.

Real Presence The conviction of the Roman Catholic Church that the body and blood of Christ are present in the bread and wine dispensed at the Mass. (For the full significance of this issue *see* Transubstantiation).

Relativity A body of theory in physics. Relativity may be said to have begun with Galileo (1564–1642), but, in practice, the word almost invariably refers to the Special and General Theories of Relativity produced by Albert Einstein. The most important innovation is that Newton's conceptions of absolute time and absolute space are 'relativized'. Time and space are revealed to be aspects of a single time-space continuum. The one absolute that remains in the system is the speed of light. Classicism is retained by the central insistence of Einsteinian relativity that all the physical laws of the universe must be the same for all observers.

Religion A way of justifying and explaining human life and its vicissitudes. The word usually refers to one of the systematic, transcendent systems like Islam, Christianity or Buddhism. I tend to employ it slightly more loosely as any way of providing meaning and purpose – environmentalism, for example – though I see that these might better be defined as substitutes for religion. Science, with its denial of meanings and purpose as scientific issues, can be seen as the opposite of religion. But the way it turns this denial into the social and ethical system of liberalism means that it behaves like a religion. It might be said to be religion's shadow, its nihilistic brother.

Science The procedures and body of knowledge that sprang from the innovations – technical and intellectual – of the sixteenth and seventeenth centuries. Clearly it is possible to refer to earlier types of science – Greek, medieval or whatever – but, for my purposes, the word is employed to refer to modern science which, as I insist throughout, is quite different in its nature and effectiveness from anything that went before.

Superstrings The products of one of the currently most successful theories in physics. Superstrings are inconceivably tiny entities that replace the preceding idea of a point particle as the most fundamental object in nature. String behaviour solves a number of mathematical problems arising from point particles.

Symmetry Symmetry is the major theme in current physics. In essence, it means what it commonly means – a matching of parts, a system in balance. To assume the existence of symmetry in the basic fabric of matter has proved an immensely powerful tool for generating theories. It is a kind of effective faith that things will fall into a pattern. Symmetry, in this view, might be said to be the natural, undisturbed condition of the universe. This condition we would call 'nothing' because symmetry would mean that all forces would be perfectly counteracted and therefore cancelled out. The existence of anything means the breaking of symmetry. We live in an enormously complex mass of broken symmetries by the close examination of which physics hopes to define the initial symmetries.

Theodicy Sometimes said to be the explanation/justification of the ways of God to man. Specifically used to deal with the problem of evil – if the world was made and is overlooked by a benevolent and omnipotent god, how can we explain the existence of evil? Science effectively answered this question by saying we could not search for value in the world. We could describe nature but we could only do so objectively and without imposing our notions of good and evil. Our theodicy would have to look elsewhere. The word is not used in the book but is included here because of the way it summarizes the core of my argument.

Theories of everything A currently fashionable phrase in physics used to describe certain types of theory which aspire to completion. Ideally, for the physicist, these would lead to a set of equations that would account for the behaviour of matter under all circumstances. Their importance is that they clearly signal the ambition of science to be *utterly* conclusive in that they go beyond the questions of What? and How? to the question Why?

Thomism The doctrine arising from the thought of St Thomas Aquinas (1225–74).

Time Considered by Newton to be an absolute, a kind of supreme clock measuring the duration of all things, but by Einstein to be a relative quantity interwoven with space. Such divisions are inevitable in any attempt to define the concept. Time can be subjective – how long we feel has passed – or objective – the movements of clocks, the processes of change and decay. Similarly it can be regarded as built into the universe or put there as a convenience by us. Either way it is now at the centre of physics as surely as it is at the centre of our consciousness.

Transubstantiation The explanation provided by St Thomas Aquinas and specifically endorsed by the Council of Trent for the mechanism whereby the Real Presence of Christ was transmitted into the bread and wine of Holy Communion. The explanation distinguished between the accidental features of matter – the smell, taste and texture of bread and wine – and their substance – the real body and blood of Christ.

Truth All things to all men. But this distinction is, for my purposes, important: there is truth in the sense of in accord with the facts of the world, or there is truth in the sense of in accord with our experience. The first can be established by the effectiveness of a particular interpretation. What the second can be established by remains and will always remain a consoling mystery.

Turing test Devised by the mathematician Alan Turing to establish whether a computer could be said to 'think'. A human interviewer speaks to a computer and another human without being able to identify which is which other than by question and answer. If he cannot tell the difference, the computer can be said to have passed.

Vacuum Generally used to mean an emptiness or, more specifically, an absence of air. Its importance is similar to that of causality in that it is a physical concept whose interpretation varies significantly in different systems. There could be no vacuum in Aristotelian physics as space was defined solely by matter. In Newtonian physics space was absolute so there could be empty space and thus a vacuum. But the vacuum vanishes again in modern physics. Quantum theory indicates that all space is replete with fields and is occupied by virtual particles which can become fully-fledged particles apparently out of nothing. So, currently, there is no such thing as a nothing.

Notes

Introduction

1 Stephen W. Hawking, *A Brief History of Time: from the big bang to black holes* (Bantam, 1988) page 175

1 Science works but is it the truth?

1 In Max Perutz, *Is Science Necessary? Essays on science and scientists.* (Oxford, 1989). Taken from the Proceedings of the National Institute of Science of India.

2 Stephen W. Hawking, *A Brief History of Time: From the big bang to black holes* (Bantam, 1988) page 175

3 Ibid, page x

4 In Max Perutz, *Is Science Necessary? Essays on science and scientists.* (Oxford, 1989) page 5

5 *Pascal's Pensées*, trans. W. F. Trotter (Everyman, 1947) pages 17–18

6 John D. Barrow *The World within the World* (Oxford, 1988) page 161

7 Freeman Dyson, *Infinite in All Directions* (Penguin, 1980) page 8

8 Ludwig Wittgenstein, *Tractatus Logico-Philosophicus* (Routledge & Kegan Paul, 1951) page 187

2 The birth of science

1 In I. Bernard Cohen, *The Birth of the New Physics* (Penguin, 1987) page 76

2 Max Weber, *The Sociology of Religion*, trans. Ephraim Fischoff (Beacon Press, Boston, 1964) page 131

3 *Pascal's Pensées*, trans W. F. Trotter (Everyman, 1947) page 65

4 In I. Bernard Cohen, *The Birth of a New Physics* (Pelican, 1987) page 56

5 In Pietro Redondi, *Galileo: Heretic* (Penguin, 1989) page 37
6 In Thomas S. Kuhn, *The Copernican Revolution: planetary astronomy in the development of western thought* (Harvard, 1970) page 126
7 Ronald W. Clark, *Einstein: The Life and Times* (Hodder & Stoughton, 1973) page 366
8 Frank E. Manuel, *A Portrait of Isaac Newton* (Frederick Muller, 1980) page 380
9 Ibid, page 85
10 Max Planck: *Where is Science Going?* trans. and ed. James Murphy (George Allen & Unwin, 1933). Quote from introduction by Albert Einstein, page 13
11 In Freeman Dyson, *Infinite in All Directions* (Penguin, 1989) page 49
12 Ibid, page 48
13 Frank E. Manuel, *A Portrait of Isaac Newton* (Frederick Muller, 1980) pages 388–9
14 Freeman Dyson, *Infinite in All Directions* (Penguin, 1989) page 50
15 In John D. Barrow, *The World within the World* (Oxford, 1988) page 30
16 Frank E. Manuel, *A Portrait of Isaac Newton* (Frederick Muller, 1980) page 119

3 The humbling of man

1 In Stephen Jay Gould, *Hen's Teeth and Horse's Toes* (Penguin, 1990) page 44
2 Bertrand Russell, *History of Western Philosophy* (Unwin, 1961), page 517
3 Frank E. Manuel, *A Portrait of Isaac Newton* (Frederick Muller, 1980) page 387
4 Alasdair MacIntyre: *Against the Self-Image of the Age: essays on ideology and philosophy* (Duckworth, 1971) pages 76–7
5 *Pascal's Pensées*, trans. W. F. Trotter (Everyman, 1947) page 23
6 John Donne, 'An Anatomy of the World: The First Anniversary' John Donne *Complete Poetry and Selected Prose* (Nonesuch, 1929) page 202
7 *Pascal's Pensées*, trans. W. F. Trotter (Everyman, 1947) page 77
8 Ibid, page 83
9 R. H. Tawney, *Religion and the Rise of Capitalism* (Penguin, 1987) page 273
10 Bernard Williams, *Descartes: the project of pure enquiry* (Penguin, 1990) page 19

11 Alexander Pope, 'Intended for Sir Isaac Newton' in *Poetical Works*, Vol. 3 (The Aldine Edition, 1934) page 142

12 In Ian Stewart, *Does God Play Dice? The Mathematics of Chaos* (Basil Blackwell, 1989) page 10

13 In Richard Dawkins, *The Blind Watchmaker* (Penguin, 1989) page 4

14 Allan Bloom, *The Closing of the American Mind* (Penguin, 1988) page 162

15 In Stephen Jay Gould, *An Urchin in the Storm* (Penguin, 1990) page 99

16 In Stephen Jay Gould, *Time's Arrow, Time's Cycle* (Pelican, 1988) page 62

17 Richard Dawkins, *The Blind Watchmaker* (Penguin, 1988), page 6

18 Arthur Schopenhauer, *The World as Will and Representation*, trans. E. F. J. Payne, Vol II (Dover, 1966) page 3

19 In Ernest Jones, *The Life and Work of Sigmund Freud* (Pelican, 1964) page 493

20 Sigmund Freud, *The Future of an Illusion* in Vol 12 of Pelican Freud Library, page 214

21 In Ronald W. Clark, *Einstein: The Life and Times* (Hodder & Stoughton, 1973) pages 48–9

22 Sigmund Freud, *Civilization and its Discontents* in Vol 12 of Pelican Freud Library, page 282

23 Ibid, page 274

24 Ibid page 271

4 Defending the faith

1 'Dover Beach' in *Matthew Arnold* (Oxford, 1986) page 136

2 Stephen Jay Gould, 'Non moral Nature' in *Hen's Teeth and Horse's Toes* (Penguin, 1990) page 35

3 Friedrich Nietzsche, *Ecce Homo* (Penguin, 1979) page 121

4 Ibid, page 127

5 Ibid, page 34

6 Ibid, page 34

7 John Keats, 'Ode on a Grecian Urn' in *Poetical Works* (Oxford, 1970) page 210

8 Max Weber, *The Sociology of Religion*, trans. Ephraim Fischoff (Beacon Press, Boston, 1964) page 22

9 Ibid, page 132

10 John D. Barrow, *The World within the World* (Oxford, 1988) page 149

11 Max Weber, *The Sociology of Religion*, trans. Ephraim Fischoff (Beacon Press, Boston, 1964) page 75

12 In *Essential Writings of Karl Marx*, sel. David Caute (MacGibbon & Kee, 1967) page 43

13 In Alasdair MacIntyre, *Against the Self-Images of the Age: Essays on ideology and philosophy* (Duckworth, 1983) page 17

14 Irving Babbitt, *Rousseau and Romanticism* (Texas, 1977) page 96

15 R. H. Tawney, *Religion and the Rise of Capitalism* (Penguin, 1987) page 55

16 Max Weber, *The Sociology of Religion*, trans. Ephraim Fischoff, (Beacon Press, Boston, 1964) page 92

17 Ibid, page 125

18 Sigmund Freud, 'The Future of an Illusion' in Vol 12 of the Pelican Freud Library, 1985, page 212

19 Ibid, page 339

20 In Stephen Jay Gould, *An Urchin in the Storm* (Penguin, 1990) page 180

5 From scientific horror to the green solution

1 James Lovelock, *Gaia: a new look at life on earth* (Oxford, 1982) page 140

2 T. S. Eliot, 'The Hollow Men', in *Collected Poems 1909–1962* (Faber & Faber, 1974) page 89

3 Max Weber, *The Sociology of Religion*, trans. Ephraim Fischoff (Beacon Press, Boston, 1964) page 117

4 Norman Stone, *Europe Transformed 1878–1919* (Fontana, 1983) page 390

5 Irving Babbitt, *Rousseau and Romanticism* (Texas, 1977) page 13

6 Wilfred Owen, 'Futility' in *The Collected Poems of Wilfred Owen* (Chatto & Windus, 1963) page 58

7 Ezra Pound, 'E. P. Ode pour l'election de son sepulchre' in *Selected Poems* (Faber & Faber, 1935) page 160

8 Hugh Thomas, *An Unfinished History of the World* (Pan, 1981) page 472

9 Max Perutz, *Is Science Necessary? Essays on science and scientists* (Oxford, 1991) pages 96–7

10 Martin Amis, *Einstein's Monsters* (Cape, 1987) page 8

11 Donella H. Meadows, Dennis L. Meadows, Jorgen Randers, William H. Behrens III: *The Limits to Growth: A report for the Club of Rome's project on the predicament of mankind* (Potomac Associates, 1972) page 141

12 Ibid, page 196
13 Bill McKibben, *The End of Nature* (Viking, 1990) page 54
14 Ibid, page 42
15 Ibid, page 195

6 A new, strange mask for science

1 In Hanbury Brown, *The Wisdom of Science: its relevance to culture and religion* (Cambridge, 1986) page 78
2 Ibid, page 66
3 P. W. Atkins, *The Creation* (W. H. Freeman & Co, 1981) page 3
4 In Ronald W. Clark, *Einstein: The Life and Times* (Hodder & Stoughton, 1973) page 73
5 Arthur Conan Doyle, *The Sign of Four* (George Newnes, 1893) page 93
6 Ronald W. Clark, *Einstein: the Life and Times* (Hodder & Stoughton, 1973) pages 227–8
7 Ibid, page 33
8 Max Planck, *Where is Science Going?*, trans. and ed. James Murphy (George Allen & Unwin, 1933) page 158
9 In Roger Penrose, *The Emperor's New Mind* (Vintage, 1990) page 361
10 In Werner Heisenberg, *Physics and Beyond: encounters and conversations*, trans. Arnold J. Pomerans (George Allen & Unwin, 1971) page 38
11 James Gleick, *Chaos: making a new science* (Heinemann, 1988) page 6

7 New wonders . . . new meanings?

1 Richard Feynman, *What do you care what other people think? Further adventures of a curious character* (Unwin, 1990) page 244
2 Ibid, page 11
3 Ibid, page 72
4 Stephen W. Hawking, *A Brief History of Time: from the big bang to black holes* (Bantam, 1988) page x
5 Freeman Dyson, *Infinite in All Directions* (Penguin, 1990) page 163
6 John D. Barrow, *Theories of Everything: the quest for ultimate explanation* (Clarendon Press, 1991) page 74
7 *Pascal's Pensées*, trans. W. F. Trotter (Everyman, 1947) page 61
8 John D. Barrow and Frank J. Tipler, *The Anthropic Cosmological Principle* (Oxford, 1988) page 677

9 Rupert Sheldrake, *The Presence of the Past: Morphic resonance and the habits of nature* (Collins, 1988) page 326
10 David Bohm, *Wholeness and the Implicate Order* (Routledge & Kegan Paul, 1980) page 138
11 Ibid, page 176
12 Max Planck, *Where is Science Going?*, trans. and ed. James Murphy (George Allen & Unwin, 1933) page 66
13 Fritjof Capra, *The Tao of Physics: an exploration of the parallels between modern physics and eastern mysticism* (Flamingo, 1990) page 28
14 Ibid, page 340

8 The assault on the self

1 Sören Kierkegaard, *The Sickness unto Death*, trans. Alastair Hannay (Penguin, 1989) page 35
2 Douglas Hofstadter, *Gödel, Escher, Bach: an eternal golden braid* (Penguin, 1980) page 709
3 Erwin Schrödinger, *What is Life? Mind and Matter* (Cambridge, 1967) page 127
4 Hannah Arendt, *The Human Condition* (Chicago, 1958) page 266
5 In Oliver Sacks, 'Neurology and the Soul' (The *New York Review of Books*, 22 November 1990) page 45
6 In Erwin Schrödinger, *What is Life? Mind and Matter* (Cambridge, 1967) page 129
7 Ibid, page 93
8 John D. Barrow, 'The Mathematical Universe' in *The World and I*, May 1989 page 311
9 Ibid, page 311
10 Douglas Hofstadter, *Gödel, Escher, Bach: an eternal golden braid* (Penguin, 1980) page 170
11 Ibid, page 710
12 Roger Penrose, *The Emperor's New Clothes: concerning computers, minds and the laws of physics* (Vintage, 1990) page 579

9 The humbling of science

1 Ludwig Wittgenstein, *Philosophical Investigations* (Blackwell, 1972) page 85
2 Max Weber, *The Sociology of Religion*, trans. Ephraim Fischoff (Beacon Press, Boston, 1964) page 137
3 Allan Bloom, *The Closing of the American Mind* (Penguin, 1989) page 26

4 Ibid, page 295

5 Ibid, page 42

6 Bertrand Russell, *The Impact of Science on Society* (Unwin Hyman, 1985) page 102

7 Ludwig Wittgenstein, *Philosophical Investigations* (Blackwell, 1972) page 42

8 Stanley Cavell, *Must we mean what we say?* (Cambridge, 1976) page 85

9 John Updike, *Self-Consciousness* (André Deutsch, 1989) page 219

10 *Pascal's Pensées*, trans. W. F. Trotter (Everyman, 1947) page 352

Bibliography

Aquinas, Thomas, *Summa Theologiae* (Blackfriars and others, London, 1964)

Arendt, Hannah, *The Human Condition* (University of Chicago Press, 1958)

Atkins, P. W., *The Creation* (W. H. Freeman & Co, 1981)

Ayer, A. J., *Ludwig Wittgenstein* (Weidenfeld & Nicolson, 1985)

Babbitt, Irving, *Rousseau and Romanticism* (University of Texas Press, 1977)

Bacon, Francis, *The Works of Francis Bacon*, collected and edited by James Spedding (Longman and others, London, 1857)

Barrow, John D., *The World Within the World* (Oxford, 1988)

Barrow, John D., *Theories of Everything: The Quest for Ultimate Explanation* (Clarendon Press, 1991)

Barrow, John D. and Tipler, Frank J., *The Anthropic Cosmological Principle* (Oxford, 1986)

Bell, Daniel, *The End of Ideology: on the exhaustion of political ideas in the fifties* (The Free Press of Glencoe, Illinois, 1960)

Bloom, Allan, *The Closing of the American Mind* (Simon and Schuster, 1987)

Boden, Margaret, A., *The Creative Mind: Myths and Mechanisms* (Weidenfeld & Nicolson, 1990)

Bohm, David, *Wholeness and the Implicate Order* (Routledge & Kegan Paul, 1980)

Bohm, David, *Causality and Chance in Modern Physics* (Routledge & Kegan Paul, 1984)

Bramwell, Anna, *Ecology in the 20th Century: A History* (Yale, 1989)

Braudel, Fernand, *On History*, trans. Sarah Matthews (Weidenfeld & Nicolson, 1980)

Bronowski, Jacob, *The Ascent of Man* (BBC, 1976)

Brown, Alan, *Modern Political Philosophy: Theories of the Just Society*, (Pelican, 1986)

Brown, Hanbury, *The Wisdom of Science: its relevance to culture and religion* (Cambridge, 1986)

Capra, Fritjof, *The Tao of Physics: An exploration of the parallels between modern physics and eastern mysticism* (Wildwood House, 1975)

Carson, Rachel, *Silent Spring* (Hamish Hamilton, 1963)

Cavell, Stanley, *Must we mean what we say?* (Cambridge, 1976)

Cellini, Benvenuto, *Autobiography*, trans. George Bull (Penguin, 1956)

Clark, Ronald, W., *Einstein: The Life and Times* (Hodder & Stoughton, 1973)

Cohen, I. Bernard, *The Birth of a New Physics* (Penguin, 1987)

Cupitt, Don, *The World to Come* (SCM Press, 1982)

Cupitt, Don, *The Leap of Reason* (Sheldon Press, 1976)

Cupitt, Don, *The Sea of Faith: Christianity in change* (BBC, 1984)

Darwin, Charles, *The Origin of Species* (John Murray, 1888)

Davies, Paul and Gribbin, John, *The Matter Myth: towards 21st-century science* (Viking, 1991)

Davis, Philip J. and Hersh, Reuben, *Descartes' Dream: The world according to mathematics* (Harcourt Brace Jovanovich, 1986)

Dawkins, Richard, *The Blind Watchmaker* (Longman, 1986)

Dawkins, Richard, *The Selfish Gene* (Oxford, 1976)

Dennett, Daniel C., *Consciousness Explained* (Little, Brown, 1991)

Descartes, René, *The Philosophical Writings of René Descartes*, Vols 1 & 2 (Cambridge, 1985)

Dyson, Freeman, *Infinite in All Directions* (Harper & Row, 1988)

Eliot, T. S., *Notes Towards the Definition of Culture* (Faber & Faber, 1948)

Feynman, Richard P., *What Do You Care What Other People Think? Further adventures of a curious character* (W. W. Norton, 1988)

Freud, Sigmund, *The Future of an Illusion*, Vol 12 The Pelican Freud Library (Pelican, 1985)

Freud, Sigmund, *Civilization and its Discontents*, Vol 12 The Pelican

Fukuyama, Francis, *The End of History and the Last Man*, (Hamish Hamilton, 1992)

Freud Library (Pelican, 1985)

Gardner, Martin, *Gardner's Whys and Wherefores* (University of Chicago Press, 1989)

Gleick, James, *Chaos: Making a New Science* (Heinemann, 1988)

Hawking, Stephen W., *A Brief History of Time: from the big bang to black holes* (Bantam, 1988)

Heisenberg, Werner, *Physics and Beyond: encounters and conversations*, trans. Arnold J. Pomerans (George Allen & Unwin, 1971)

Hofstadter, Douglas, *Gödel, Escher, Bach: an eternal golden braid* (Penguin, 1980)

Howard, Jonathan, *Darwin* (Oxford, 1982)

Huxley, Aldous, *Literature and Science* (Chatto & Windus, 1963)

Gould, Stephen Jay, *Hen's Teeth and Horse's Toes: further reflections in natural history* (W. W. Norton, 1983)

Gould, Stephen Jay, *An Urchin in the Storm* (William Collins & Sons, 1988)

Jones, Ernest, *The Life and Work of Sigmund Freud* (Hogarth Press, 1957)

Kant, Immanuel, *The Critique of Pure Reason* (Everyman, 1940)

Kenny, Anthony, ed., *Rationalism, Empiricism and Idealism: British Academy Lectures on the History of Philosophy* (Clarendon Press, 1986)

Kierkegaard, Sören, *The Sickness unto Death*, trans. Alastair Hannay (Penguin, 1989)

Kierkegaard, Sören, *Either/Or* Vols I and II, trans. David F. Swenson and Lillian Marvin Swenson (Princeton, 1971)

Kuhn, Thomas S., *The Copernican Revolution: planetary astronomy in the development of western thought* (Harvard, 1970)

Lovelock, James, *Gaia: a new look at life on earth* (Oxford, 1979)

Lovelock, James, *The Ages of Gaia: a biography of our living earth* (Oxford, 1988)

Meadows, Donella H.; Meadows, Dennis L.; Randers, Jorgen; Behrens, William W., *The Limits to Growth: A Report for the Club of Rome's Project on the Predicament of Mankind* (Potomac Associates, 1972)

Macintyre, Alasdair, *Against the Self-Images of the Age; essays on ideology and philosophy* (Duckworth, 1971)

Macintyre, Alasdair, *After Virtue a study in moral theory (Duckworth, 1981)*

McKibben, Bill, *The End of Nature* (Viking, 1990)

Manuel, Frank E., *A Portrait of Isaac Newton* (Frederick Muller, 1980)

Medawar, Peter, *The Limits of Science* (Oxford, 1985)

Monk, Ray, *Ludwig Wittgenstein: The Duty of Genius* (Jonathan Cape, 1990)

Mumford, Lewis, *The Urban Prospect* (Secker & Warburg, 1968)

Nietzsche, Friedrich, *Ecce Homo* (Penguin, 1979)

Pascal, Blaise, *Pascal's Pensées*, trans. W. F. Trotter, introduction by T. S. Eliot (Everyman, 1947)

Perutz, Max, *Is Science Necessary? Essays on science and scientists* (Barrie & Jenkins, 1989)

Penrose, Roger, *The Emperor's New Mind: concerning computers, minds and the laws of physics* (Oxford, 1989)

Planck, Max, *Where is Science Going?*, trans. and ed. James Murphy (George Allen & Unwin, 1933)

Polkinghorne, J. C., *The Quantum World* (Longman, 1984)

Russell, Bertrand, *History of Western Philosophy* (Unwin, 1946)

Russell, Bertrand, *The Impact of Science on Society* (Unwin, 1952)

Schopenhauer, Arthur, *The World as Will and Representation*, Vols I & II, trans. E. F. J. Payne (Dover, 1966)

Schrödinger, Erwin, *What is Life? Mind and Matter* (Cambridge, 1967)

Sheldrake, Rupert, *The Presence of the Past: morphic resonance and the habits of nature* (Collins, 1988)

Stone, Norman, *Europe Transformed 1878–1919* (Fontana, 1983)

Redondi, Pietro, *Galileo: Heretic*, trans. Raymond Rosenthal (Princeton, 1987)

Stewart, Ian, *Does God Play Dice? The mathematics of chaos* (Basil Blackwell, 1989)

Strauss, David Friedrich, *A New Life of Jesus* (Williams and Norgate, 1879)

Tawney, R. H., *Religion and the Rise of Capitalism* (John Murray, 1926)

Thomas, Hugh, *An Unfinished History of the World* (Hamish Hamilton, 1979)

Weber, Max, *The Sociology of Religion*, trans. Ephraim Fischoff (Beacon Press, Boston, 1964)

Williams, Bernard, *Descartes: the project of pure enquiry* (Pelican, 1978)

Wittgenstein, Ludwig, *On Certainty*, trans. Denis Paul and G. E. M. Anscombe (Basil Blackwell, 1969)

Wittgenstein, Ludwig, *Philosophical Investigations*, trans. G. E. M. Anscombe (Basil Blackwell, 1972)

Zohar, Danah, *The Quantum Self* (Bloomsbury, 1990)

Index

Absolute space 35, 41, 68, 153
Absolute time 35, 41, 68, 153
Abu Ja'far Muhammad ibn Musa
    al-Khowarizm 43
Acid rain 133
Adams, Douglas 177
Adolescence 235–6
Adrian, Lord Edgar Douglas 208
Affluence 124
Agriculture: chemicals used in
    127–9; cycles of 85
AI (artificial intelligence) 220–21,
    252; *see also* Hard AI
Air resistance 149
Alexander the Great 23–4
Alien 172
Alienation 93–4
Aliens 172
'Algorithmic compression' 43,
    182–3, 197, 221
Algorithms 219, 221, 222, 223,
    224
Almagest 25
Amis, Martin 122
Anglo-Dutch conflict 56
'Angular unconformities' 71
Anthropic Principle 186–7, 196,
    247, 251–2; *see also* Fridge-
    Light Hypothesis
'Anthropic selection effect' 184
Anthropology 200
Anti-progressive movements 126
Apocalypse 123, 133, 136
Apocalypse Now 113
Aquinas, Saint Thomas 18, 19,
    21–3, 26, 28, 30, 36, 38, 40,
    52, 54, 62, 67, 88, 105, 143
Arab genius 33
Arabic numerals 33
Architecture 21–2, 28
Arendt, Hannah 208
Arguments for and against science
    233–4
Aristotle 18, 19, 20, 21, 23, 24–5,
    26, 27, 28, 30, 32, 36, 37, 42,
    51, 62, 63, 67, 87, 104, 159,
    180; works of 22
Aristotelianism 25, 27, 28, 31–2,
    36, 38, 43, 50–51, 52, 60, 61,
    65, 73, 88, 186, 252; *see also*
    Thomism
Arnold, Matthew 79, 108–9
Art and literature: modern
    imagination in 112–13;
    modernistic works of 170
Astrology/astronomy, Aristotelian
    theory 24–5
Astronomy 24–5, 169; Ptolemy's
    system 23, 25
Atkins, Peter 150–51, 208
Atom bomb 114, 121–3, 139, 140,
    171, 228
Auschwitz 120, 122
'Authentic choice' 100, 252
Autonomous vision of existence
    179–80, 220
Ayatollah Khomeini 13
Ayer, A. J. 230

Babbage, Charles 218
Babbitt, Irving 103, 118

273

Bacon, Francis 50–52, 81
Barberini, Cardinal Maffeo 37
Barrow, John 16, 89, 182, 187, 219
Beckett, Samuel 59–60
Beethoven, Ludwig van 120, 155, 244
Being and Nothingness 58
Bible, the 133; non-literal interpretations of 97–8; primacy of 55
Bicycle, the 115
Big Bang 179, 180–2, 184, 185, 186, 252
Big Crunch 180, 253
Billiard balls 143, 145, 146, 147, 153, 178, 196
Biology 188–9, 204
Black hole 253
Bloom, Allan 67, 237–9, 246, 248
Boabdil, last Arab King of Spain 33
Bohm, David 189–92, 193, 196, 230
Bohr, Niels 122, 138, 152, 154, 158, 161, 194
Book of Faith 90–91
Book of Nature 62, 90–91
Bourgeois, the 104–6, 107, 108, 126
Brahe, Tycho 37
Bramwell, Anna 125
Brief History of Time, A 1, 178, 179
Bronowski, Jacob 2, 4, 230, 247, 248
Buddhism 86, 193, 194
'Butterfly effect' 161, 166

Caesar, Julius 117
Calculators 218, 220
Calculus 253; *see also* Leibniz, Gottfried *and* Newton, Isaac
Calvin, John 32
Calvinism 202
Cantor, Georg 216
Capra, Fritjof 192–4, 196, 230

Carroll, Lewis 15, 93, 153, 222
Carson, Rachel 126–9, 136
Cartesianism 50, 52, 53, 58, 59, 65, 75, 137, 159, 175, 208, 210, 242, 244, 245, 253; *see also* Descartes, René
Cassiopeia 37
Categorical imperative 253; *see also* Kant, Immanuel
Cathedral of Chartres 22
Causality 26–7, 76, 83, 95, 163, 178, 186, 188, 232, 253
Cavell, Stanley 243
Cellini, Benvenuto 61
CFCs 111, 127
Change: idea of 235–7; of emphasis 247
Changes in science 141
Chaos/chaology 141, 151, 161–6, 171, 177, 178, 194, 230, 253
Chemicals used in agriculture 127–9
Chicago 148–9, 162
Chinese religion, holistic 89
Christian and classical wisdom, need to unite 21–2
Christian eschatology 201
Christian Europe, effect of science on 90–91
Christianity 20–21, 23, 51, 70, 86–8, 98–9, 100, 173, 191, 201
Churchill, Sir Winston 13–14
Classicism 25, 27, 62, 135, 140–41, 145, 148–51, 154–5, 157–9, 160–61, 162, 164, 168, 170, 180–81, 183, 185, 186, 187, 189–93, 196, 202, 206, 211, 215, 216, 223, 242, 253–4
Clear Lake, California 127–9, 136, 140
Clocks, invention of 34–5
Close Encounters of the Third Kind 172
Closing of the American Mind, The 237
Club of Rome report 1972 130–32
Cobalt Bomb 123

Cogito, ergo sum 52, 57, 77, 241, 244, 254
Cognizance, subject of 208–9
'Coincidence' 184
Columbus, Christopher 29, 30, 33, 34, 36
Communications 115, 211
Compass bearings 34
Complementarity 158
Completion 142–3, 146–7, 166, 177, 217; ambition for 179–80
'Complex chaos' 166
Complex systems, explanations of 66–7
Compromise between nature and science 154
Computers 173–5, 218–19, 220–21, 224, 247; chess-playing 221; processing power of 225
Confucianism 86, 89
Conrad, Joseph 112–13, 116
Consciousness 206–7, 215, 219, 224, 244
Consciousness Explained 245
Controllable measurement process 162
Copernican-Keplerian system 41
Copernicanism 31–2, 38, 132, 254
Copernicus, Nicolaus 31–2, 36, 63, 76, 187, 190
Coppola, Francis Ford 113
Cornell University 166
Corruption, Christian and scientific 191
Cosmological constant 180, 254
Cosmology 20, 22, 23–4, 25, 56, 63, 68, 84, 90, 107, 143, 156, 179, 197, 204, 205, 237, 254
Council of Trent 1545 39, 55
Counter-Reformation 28, 55, 91, 201
Creation, The 151
Crick, F. H. C. 74–5
Critique of Pure Reason 67
Cuba 139; missile crisis 123
Cultural liberalism 97
Cultural pessimism 112, 117–18

Culture 7–9, 52, 79–81, 92, 94, 103, 177, 189, 191, 195, 228, 232, 235, 236, 237–9, 240, 245, 250, 254

Dachau 139
Dampier, Sir William 75
Dante, Alighieri 18, 19, 36, 153
Darwin, Charles 48, 72–3, 75–7, 80, 84, 108, 113–14, 134, 142, 167, 200
Dawkins, Richard 73, 178, 206, 246
Death, modern 118
DDD pesticide 127–9
DDT pesticide 127, 136
Debating points 233–5
Decadence, terminal 248–9
de Chardin, Teilhard 97
'Deep ecology' 136
'Deep time', concept of 71–2
'Democratic values' 113
Dennett, Daniel C. 245
De revolutionibus orbium coelestium 31
Descartes, René 39, 47, 49–50, 51, 52, 53, 56–60, 64, 65, 67, 69, 77, 79, 90, 91, 94, 114, 153, 158, 175, 203, 209–10, 227, 240, 241–2, 243, 246; *see also* Cartesianism
Design, argument from 73, 254; *see also* God
Determinism 64, 162, 172, 178, 195–6, 203, 254–5
Dicke, Robert 184, 186
Dirac, Paul 184, 186, 187, 211
Discourse de la Méthode . . . 39, 49, 52, 57, 59
Divine Comedy 19, 20
DNA 74, 215, 255
Dominicans 22
Donne, John 53
Dualism 126, 210
Dyson, Freeman 16, 45, 180

East, the, and science 88–9

Eclipse 155
Ecology 125–6, 128, 132–3, 135;
    Gaia hypothesis 129–30, 136
Economics 115 -16; energy 125
Eddington, Arthur 155, 169
Effectiveness of science 5–9
Einstein, Albert 1, 15, 37, 42, 57,
    65, 88, 120, 122, 123, 153–6,
    159–60, 163, 169, 179–80, 195,
    196, 197, 204, 211
Einstein, Rosen and Podolsky
    'thought' experiment 160
Electricity 115, 143–5
Electromagnetic phenomena
    143–5
Electronics 173–5, 218; *see also*
    Computers
Eliot, George 98
Eliot, T. S. 112, 116, 170
Ellis, Havelock 75
Embryos, experiments on human
    233–4
Emperor's New Mind, The 221
Empiricism 67, 255; *see also*
    Rationalism
*End of History and the Last Man,
    The* 10, 232
End of Nature, The 134
Energy economics 125
Enlightenment 67, 82, 101, 114,
    116, 118, 120, 122, 139,
    140–41, 167, 181, 187, 191,
    192, 196, 203, 214, 230, 234,
    235, 242, 245, 248, 249
Environment, Department of the
    111
Environmentalism 110–11,
    124–6, 129–37, 228–9, 247
Epistemology 58, 67, 82, 90, 158,
    181, 255
ET 74, 172, 174
Ether 143, 145–6
'Ether wind' 146
Euclid 62, 216
Evolution 188, 222
Evolutionary complexity 206

Existentialism 58, 101, 129–30,
    255
Exponential growth 131

Faith 179, 216, 227, 229, 234,
    242; 'authentic choice'
    100–101; in environmentalism
    124, 132–3, 135; book of 90–91;
    meaning of 99–100
Faraday, Michael 143–5, 151, 200
Fascism 126
Ferdinand and Isabella, King and
    Queen of Spain 33; motto of 29
Feuerbach, Ludwig (Luther II)
    93–4
Feynman, Richard 168–70, 177,
    194, 230
Field theories 156, 179, 182, 255
Final Anthropic Principle 186–7
First Machine Age 117
First World War 114, 117
Flaubert, Gustave 102–4, 105,
    108
Fragmentation 192–3
Franciscans 22, 23, 88
Frankenstein 171
Franklin, Benjamin 69
Frege, Gottlob 216
Freud, Sigmund 75–8, 80, 84,
    107–8, 114, 167, 200, 204,
    205–6, 215
Fridge-light Hypothesis 185–6,
    255; *see also* Anthropic
    Principle *and* Quantum Theory
Fukuyama, Francis 10, 14,
    232–3

Gadaffi, Colonel 33
Gaia hypothesis 129–30, 136
Galen 30
Galilean-Newtonian system 143
Galileo, Galilei 17–19, 23, 24, 30,
    32, 35, 36, 37, 38, 39, 42, 46–7,
    48–9, 53, 54, 56, 58, 59, 63, 67,
    69, 70, 71, 78, 88, 90, 93, 114,
    117, 147, 149–50, 152, 162,
    166, 190, 209, 240

'Gauge theories' 182
Geocentrism 84, 255; *see also*
	Heliocentrism
Geology as a science 71
Geometry 165, 216
Global equilibrium 131–2
Global solutions, search for 115
God 57, 99, 101, 125, 133, 159,
	160, 179, 183, 202–3, 210, 242,
	246, 255; all-powerful 89; and
	classicism 159–60; as organizer
	of complexity 73; as structure
	of self 57, 59; definition of 40;
	evocation of 2–3, 16; existence
	of 100; position of in world 60,
	61, 84; pre-scientific 125;
	science and 9, 12;
	understanding 65–6, 179
Godel, Kurt 217, 218, 219, 221
*Godel, Escher Bach: an eternal
	golden braid* 221
Godel's incompleteness theorem
	217–18, 255–6
Gothic cathedrals 21–2
Gould, Stephen Jay 80–81
Gravity 143–4, 148, 149, 153–4,
	206
'Gravity shear' 187
Great Instauration, The 50
'Green' impulse 111, 124, 126,
	130, 133, 137, 231, 245
Green Movement 125, 129
Greenhouse Effect 127
Growth and prosperity, period of
	116
Guernica 120
Gulliver's Travels 171
Gustavus Adolphus, King of
	Sweden 39

Haldane, J. B. S. 107
Halley, Edmund 42
Halley's Comet 92, 155
'Hard AI' 221, 223–5
'Hard evidence' 114
Hard science 228

Harmony and interrelatedness'
	194
Harvard University 117
Hawking, Stephen 1–2, 9, 12, 16,
	21, 142, 178–9, 230, 247, 248
Heart of Darkness 112–13
Hegel, Georg Wilhelm Friedrich
	93–4, 97, 98, 100
Hegelianism 93–4, 95, 97, 176,
	224
Heisenberg, Werner 157, 192,
	194, 195
Heliocentrism 31–2, 63, 256; *see
	also* Geocentrism
Hilbert, David 217–18
Hinduism 86, 193
Hiroshima 114, 121, 122, 139,
	140
History 93–6, 237, 239–40, 244
Hitler, Adolf 119
Hofstadter, Douglas 206–7,
	221–3, 225, 230, 245
Holistic ideals 89, 125, 193
Hollow Men, The 112
Holocaust 119
Horror and anxiety of science
	112–24, 138–9
Hubristic humanism 118
Human body as part of objective
	world 209–10
Human life, meaning and purpose
	of 227, 230, 235–7, 240;
	experiments on embryos 233
Hume, David 67, 70, 79, 90, 93,
	193
Hussein, Saddam 7, 33
Hutton, James 71
Huygens, Christian 44–5

Ichneumon wasp 80–81, 84
Idealism 82, 256
Imaginative developments 170–77
Impossible, the, enters human
	affairs 17–18
Improved mental selfhood 211
Individualism 61, 91, 202–3;
	Protestant 100

Industrialization 117, 120–21,
124; *see also* Mechanization
'Innate tendency', concept of 27
Inner fabric of science 140
Institutionalization 236, 238
Intellectualism 23, 106–7, 231
Internalization, Protestant 203
Iran, and the Salman Rushdie
affair 12–13
Islam 86
Islamic fundamentalism and
Western liberalism 12

James, Henry 115
Jesuits 39, 51, 55, 58, 91, 147
Judaism 85, 86
Jung, Carl 208, 212
Jupiter 147

Kaiser Wilhelm 117
Kant, Immanuel 67–70, 79, 82–4,
90, 93, 98, 120, 192, 193, 203,
240, 245, 248
Keats, John 84
Kelvin, Lord William Thomson
116, 142
Kennedy, President John F. 123,
176
Kepler, Johannes 17, 31, 35, 41,
42, 63, 190
Keynes, John Maynard 45–6
Khrushchev, Nikita 123
Kierkegaard, Sören 100–101,
103–4, 105, 108, 199, 202–4,
211, 245, 248
Knowledge: forms of 195, 199,
200, 208, 218, 227, 239;
pursuit of 138–9, 141–2, 190,
194, 201; and value, severance
of 113, 139
Kremer, Gerhard *see* Mercator,
Gerardus
Kubrick, Stanley 176

Language 207, 208, 217, 222,
241, 244–5; private 241
Lao-Tzu 194

Laplace, Pierre Simon, Marquis
de 64
Laplacean determinism 64, 162
Large Number Coincidence 184,
186
Laws of Motion 188
Leibniz, Gottfried Wilhelm 60,
82, 120
Lenin, Vladimir Ilich 10, 115
Leo XIII, Pope 92
Leonardo da Vinci 61
Lessing, Gotthold Ephraim 98
Liberalism 11–14, 93, 96–8, 112,
224–5, 229, 232–5, 238,
239–41, 243, 248–9, 256
*Life of Jesus, critically examined,
The* 98
Life, rhythms of 85
Light 143–4, 145–6, 158, 206;
wavelength of 231
Limits to Growth, The 130
Lincoln, Abraham 236
*Lives of the most excellent Italian
Architects, Painters and
Sculptors, The* 61
Locke, John 41
Lorentz, Hendrick Antoon 145,
151, 163
Lovelock, James 110, 129–30,
133
Low-energy 111
Luther, Martin 28, 32, 56, 82

McCarthy, John 240
Machine age 69; *see also*
Mechanization
Macintyre, Alasdair 50
McKibben, Bill 135–6
Madame Bovary 102–3, 105
Magic 92, 173, 245
Magnetism 143–5
Malaria 136
Mandelbrot Set 166
Manuel, Frank E. 40
Maps, compilation of 5–7, 34, 165
Marne, Battle of the 117

Marx, Karl 10, 94–6, 105, 115, 205–6
Marxism 94–6, 126, 235
Mass communications 211
Material benefits 113
Material optimism 114
Mathematics 33, 62–3, 150, 164, 165, 211, 216–20, 221, 222, 223
Maxwell, James Clerk 144–5, 151, 199
Maxwellian equations 145
Mechanistic science 114, 170, 178, 183, 191, 195, 211, 215, 223, 247; *see also* Newtonian Mechanics
Mechanization 115, 117, 121, 138
Medieval cosmology 25
Medieval humanism 21–2
Memory 188
Mercator, Gerardus 34, 164
Metaphysics 183
Meteorology 164
Michelangelo 48, 61
Michelson, Albert 142, 146
Middle Ages 28, 33, 35, 67, 84, 87
Modern death 118
Modernism 112
Modernist Movement in the arts 170
Monotheistic culture 51–2
Moon: discovery of 37; landing on 176
Moore, Christopher 166
Morality 68, 73, 80–81, 91, 127, 133, 135, 139, 213–14, 233–4, 236, 240
Morley, Edward 146
'Morphogenetic fields' 188–9, 196
Moses 85, 86
Motion, Laws of 188
Muhammad 86
Music 221

Nagasaki 114
Napoleon Bonaparte 64, 117

Narcissism 211–13
NASA, X-ray missions 180
Natural History Museum 81, 102
Natural selection 72, 148
Nature: autonomy of 124–5; Book of 62, 90–91; compromise with science 154; exploitation of 134; fatalism of 134; opposition to science 212–13; preservation of 129
Nazi 'science' 120
Nazism 119, 126
Nehru, Jawaharlal 1, 3–4
New Pagan selfhood 225
New science 168–9, 177, 178, 179; *see also Scienza Nuova*
New World, discovery of 28–30
Newton, Isaac 1, 23, 32, 35, 39–47, 48–9, 50, 56, 57, 59, 60–62, 63, 64, 65–6, 67, 68, 69, 70, 73, 76, 77, 80, 83, 88, 89, 90, 91, 108, 114, 117, 142, 143, 145, 146–8, 150, 151, 152–3, 154, 155, 156, 162, 179, 183, 188, 195, 204, 236–7, 240
Newtonian mechanics 40, 42, 45–6, 49, 61–2, 94, 121, 143, 144, 145, 146, 147, 148, 150, 151, 153, 154, 155, 160–61, 162, 179, 200, 256
Nietzsche, Friedrich 81–3, 94, 99, 100, 105, 204, 211–12
Nietzschean theory 82–3, 118, 119
Nineteenth century science and technology 195
'No-boundary condition' 179
Nonmoral Nature 80
Non-western cultures, impact of science on 7–9
'Noosphere' 97
Nuclear weapons 122–3, 138, 139
Numbers 215–16

'Oil shock' 132
Omega Point 187, 189
On Fabric of the Human Body 30

*On the Electrodynamics of Moving
  Bodies* 153
Ontology 205
Organic complexity 73
Orthodoxy vs science 37
Owen, Wilfred 118
Ozone layer 127, 136–7

Pacific War 122
Paley, William 66, 70
Pascal, Blaise 15, 29, 50, 53, 54,
  58, 93, 185, 248, 249
'PC'-politically correct 238
Penrose, Professor Roger 221,
  223–5, 243
Penzias, Arno 181
Perutz, Max 4, 120
Pessimism 112, 117–18, 132–3,
  168, 178, 205, 228
Pesticides 127–9, 133, 136, 139,
  170
Petrarch, Francesco 37–8
*Philosophical Essays on
  Probabilities* 64
Philosophical Investigations 241
Philosophy 221
Physical self-cultivation 211–12
Physical systems QED 225
Physics 1, 22, 24, 56, 75, 116–17,
  153, 154, 155, 156, 179, 216,
  221, 222, 237, 256–7; classical
  161, 166; new 121, 143–4, 146,
  171, 181, 183, 193
Pi, value of 62
Pillars of Hercules 29
Pineal gland 59
Planck, Max 44, 151–2, 156, 157,
  159, 161, 164, 192, 195
Planck's Constant 151–2, 156,
  157, 161
Plato 62, 63, 68
Platonism 62, 63, 87, 223, 224
Playfair, John 71
Plurality and tolerance 248–9
Poincaré, Henri 163
Politics linked with science 94–5
Pollution 111, 133, 136

Pope, Alexander 60, 65
Popularizers of science 2–3, 247,
  249
Pound, Ezra 118
Pre-Reformation Catholicism 39
Primum Mobile 24
Principia Mathematica 217
Principle, statement of 233
Progress, idea of 236–7
Protestantism 19, 39, 55, 56, 82,
  91, 92, 93, 97, 99, 173, 202–3,
  245
Providentissimus Deus 92
Pseudo-scientific metaphysics 115
Psychiatry 200
Psychoanalysis 75–7, 200, 204,
  211, 214, 231, 246, 257
Psychology 200
Ptolemaic method 66, 197
Ptolemy 23, 25, 29, 30, 31, 36,
  64, 147, 197; works of 22
Public image of science, change
  in 138
Purism 135
Pythagoras 63, 216

Quandary of the intellectual 231
Quantum electrodynamics 168–9
Quantum flux 180
'Quantum leaps' 152
Quantum theory/mechanics 54,
  116, 141, 151, 152–3, 154,
  156–61, 162, 166, 171, 177–8,
  180, 181, 189, 190, 192, 193,
  194, 196, 199, 201, 208,
  210–11, 215, 221, 230, 257
Quantum gravity theory 224

Radiation 157
Radicalism, advent of vs science
  87–8, 195
Radio 200
Raphael 61
Rationalism 104, 202, 257
Rationality 209, 213; of modern
  war 118–20, 122–3

Real Presence 201, 257; *see also* Transubstantiation
'Real' world 208
Realization 244–5, 247, 249
Recycling products 110–11
Redondi, Pietro 38
Reformation 28, 32, 39, 52, 55, 92, 202
Relativism 237, 242, 248
Relativity 54, 169, 177–8, 181, 194, 221, 258; General Theory of 42, 153, 155; Special Theory of 153; theory of 1, 38, 141, 151, 154, 161, 162, 201
Religion 107, 134, 136, 178–9, 199, 201–2, 245, 258; attitude to 55; Chinese 89; decline in power of 227; return to orthodox 229; science as 228; self sustained by 232; undermined by science 192; unity of 90–91; vs science 81–6, 91–3, 107–8
Renaissance 28, 31, 33, 37, 51, 52, 61, 104, 202; Catholicism 38
Richardson, Lewis Fry 164
Rigveda 86
Roman Catholic Church 91–3; crisis of authority 58–9; knowledge and wisdom of 53–4
Roman Catholicism 19–20, 32, 55, 87, 199, 201, 202
Romanticism 68, 124–5, 169, 204, 211
Romer, Ole 144
Rousseau, Jean Jacques 130
Royal Astronomical Society 155
Royal Road to the Truth 154
Royal Society 155
Rushdie, Salman 12–13
Russell, Bertrand 48, 79, 216–17, 230, 239–41, 242, 244, 247
Russian Revolution 115
Rutherford, Ernest 122

Sagan, Carl 2, 179, 230, 247

St Augustine 23
St Paul 54, 88
Sarpi, Paola 55
Sartre, Jean Paul 58
Satanic Verses, The 12
Scepticism 54–5, 57–9, 65, 92, 94, 105, 227, 230
Schleiermacher 98
Schopenhauer, Arthur 73
Schrödinger, Erwin 152, 208–9, 210, 212
Schweitzer, Albert 99
Science 258; advent of 30; redefining of 192–3
Science fiction 171–4, 176–7
Scientific knowledge and human happiness, discontinuity between 37–8
Scientific legacy 237
Scientific research, wartime 120
Scientific society, new values 112
Scientific solution to problems 4
'scientific' sociology and politics 116
Scientific technology 125
Scientific universe, primary elements 61–2
scientism 2
Scienza Nuova 30, 32, 36, 37, 39, 49, 52, 54, 56, 59, 61, 63, 69, 87, 216; *see also* New science
Second World War 114, 118, 126
'seeing' 223, 243
'selection effects' 185
Self: -awareness 59, 203, 206, 231; -consciousness 203, 206–8, 210, 221–3, 224, 244; -cultivation, physical 211–14; delicate structure of 57; -examination 204, 231; exclusion of, from explanations of science 14–15; exclusion of, from world 203, 208, 214; fulfilment of 201–2; glorification of 211; inner 246; -interest 234; -knowledge 205; materialistic conception of 245;

peace within, 243; -perception 223; problem of 223; the real 245; rediscovered 204; -reference 217; reintegrated sense of 211; as safe refuge 204, 214, 227; scientific 200, 205, 242; search for in numbers 215; selfhood 225, 242
Serpentarius 36
Set theory 216–17
Shakespeare, William 48, 90
Sheldrake, Rupert 188–9, 196
Shelley, Mary 171
Siddhárth 86
Sign of Four, The 152
Silent Spring 126–9
Simplifying instincts of science 128, 150–52, 156, 162, 166
Sociology 116, 200
Socrates 24
'Something missing' 223–4
Somme, Battle of 117
Spiritual condition of scientific liberalism 11–14
Spiritual impoverishment 232, 240
Spirituality, new, of science 230
Stalinist Terror 96
Star Trek 172
Star Wars 172
Statistics 163
Stone, Norman 116
Strauss, David Friedrich 98, 108
Strong Anthropic Principle 186
Subject of Cognizance 208–9
'Subjective certainty' 239
Summa Theologia 21, 22, 23, 90
'Supergravity' theory 182
Supermarkets 110–11
Superstrings 182, 197, 258
Swift, Jonathan 15, 93, 153, 171
Symmetry 70, 182, 183, 187, 192, 258–9

Taoism 193, 194, 196
Tartaglia, Niccola 30
Tawney, R. H. 56, 104

Technology 51, 69, 101, 113; and war 117–23; wonder of 171; benign 126; developments 33–5; downgrading of 177, 247; electronics 173–5; in the twentieth century 114–24; science fiction 171–2, 174, 176, 177
Telescope, the 17–18; invention of 28, 35–7
Television 200
Theodicy 259
Theology 183, 232; liberalism in 97
Theories of Everything 1, 21, 41, 54, 85, 87, 156, 182, 197, 259
Theory of types 217
Thomas, Hugh 21, 120
Thomism 23, 28, 32, 38–9, 88, 104, 143, 209, 259
'Thought' experiment 160
Tillich, Paul 97–9
Time 236, 259
Tipler, Frank 187
Tokyo 148–9, 162
Tolerance *see* Liberalism
Total Symmetry 187
Traditional science 198
Transubstantiation 38–9, 55, 84, 201, 259; *see also* Real Presence
Treaty of Westphalia 1648 56
Trevithick, Richard 69
Trowbridge, John 116
Truth 49, 56, 141–2, 146, 147–8, 153, 154, 162, 176, 195, 197, 199, 208, 228, 230, 237, 260; ultimate 18–19
Turing, Alan 218–19, 221, 222
Turing test 218–19, 222, 260
Twentieth-century science 114–24, 139–41, 156, 161, 178, 194, 195
2001: A Space Odyssey 176, 194
Typewriter, the 115

Ulm 47
Undermining authority 28–9

'Unified field theory' 156, 179
University of Tokyo 62
Unselfishness 250
Upanishads 86
Updike, John 245–6
Urban VIII, Pope 37, 39
Usury 104–5

Vacuum 14, 113, 145, 159, 260
Value distinctions 238, 240
Van Gogh, Vincent 148
Vasari, Giorgio 61
Verne, Jules 171, 172
Vesalius, Andreas 30
Virgil 20
'Vortex of self-perception' 223

War and science 117–23, 138–9
War Games 174
Waste Land, The 170
Watch, image of 66, 70
Watson, J. D. 74–5
'Weak Anthropic Principle'
    184–5, 187
Weather forecasting 163–5

*Weather Prediction by Numerical
    Process* 164
Weber, Max 23, 86–7, 92, 105,
    106, 114, 173, 200, 231

Wedgwood, Josiah 69
Weird Science 172, 174
Wells, H. G. 171, 172, 174
Whitehead, Alfred North 155,
    217
Whole Truth, The 141–2
Whoroscope 59
Wilde, Oscar 204
Wildlife, effect of chemicals on
    127–8
Williams, Bernard 57
Wilson, Robert 181
Wittgenstein, Ludwig xiii, 16, 46,
    227, 241–6, 248, 249
Wordsworth, William 130
'world machine' 71

Ypres, Battle of 117

Zoroastrianism 86